Corrosion of Steel in Conc

Corrosion of Steel in Concrete: Understanding, Investigation and Repair is a guide for designing, constructing and maintaining reinforced concrete structures, such as buildings and bridges which are subject to reinforcement corrosion. It presents the basics of theory and practice in steel corrosion in concrete and reviews the latest research and developments, such as progress on measuring the corrosion threshold for chloride-induced corrosion.

This third edition compares the currently proliferating major national and international standards and guidance documents. New developments are considered, such as hybrid anodes for electrochemical treatment and the latest research and developments in assessment, such as the use of ground penetrating radar to measure the chloride content of the concrete cover. It overhauls coverage of electrochemical repair and rehabilitation techniques and outlines recent innovations in structural repair and construction and investigates their implications for durability.

The book is ideal for practitioners and graduate students in structural engineering and concrete technology.

RUST'S A MUST

Mighty ships upon the ocean
Suffer from severe corrosion,
Even those that stay at dockside
Are rapidly becoming oxide
Alas that piling in the sea
is mostly Fe_2O_3,
And where the ocean meets the shore
You'll find there's Fe_3O_4,
'Cause when the wind is salt and gusty
Things are getting awful rusty.

We can measure we can test it,
We can halt it or arrest it,
We can gather it and weigh it,
We can coat it, we can spray it,
We examine and dissect it,
We cathodically protect it,
We can pick it up and drop it,
But heaven knows, we'll never stop it.

So here's to rust, no doubt about it,
Most of us would starve without it.

T.R.B Watson, June 1974
'Corrosion Control Services'

Corrosion of Steel in Concrete

Understanding, Investigation and Repair

Third Edition

John P. Broomfield

CRC Press
Taylor & Francis Group
Boca Raton London New York

Third edition published 2023
by CRC Press
4 Park Square, Milton Park, Abingdon, Oxon, OX14 4RN

and by CRC Press
6000 Broken Sound Parkway NW, Suite 300, Boca Raton, FL 33487-2742

© 2023 John P. Broomfield

First edition published by E & FN Spon 1997
Second edition published by CRC Press 2007

CRC Press is an imprint of Informa UK Limited

British Library Cataloguing-in-Publication Data
A catalogue record for this book is available from the British Library

Library of Congress Cataloging-in-Publication Data
Names: Broomfield, John P., author.
Title: Corrosion of steel in concrete : understanding, investigation and repair / John P. Broomfield.
Description: Third edition. | Abingdon, Oxon ; Boca Raton, FL : CRC Press, 2023. |
Includes bibliographical references and index.
Identifiers: LCCN 2022034158 | ISBN 9781032120980 (hardback) |
ISBN 9781032120997 (paperback) | ISBN 9781003223016 (ebook)
Subjects: LCSH: Reinforcing bars–Corrosion. |
Steel–Corrosion. | Reinforced concrete–Corrosion.
Classification: LCC TA445.5 .B76 2023 | DDC 620.1/37–dc23/eng/20221011
LC record available at https://lccn.loc.gov/2022034158

ISBN: 978-1-032-12098-0 (hbk)
ISBN: 978-1-032-12099-7 (pbk)
ISBN: 978-1-003-22301-6 (ebk)

DOI: 10.1201/9781003223016

Typeset in Sabon
by Newgen Publishing UK

Contents

Preface

This third edition provides updated information on techniques, materials, standards and the technical literature relating to corrosion of steel in concrete in atmospherically exposed concrete structures. There is also additional information on early 20th century steel framed stone and brick buildings. The sections on corrosion theory, corrosion monitoring, galvanic cathodic protection and other areas have been introduced or expanded to reflect the changes of the past decade since the previous edition. The increasing number of relevant national and international standards is also reflected in the text. In this edition I have also tried to cover some relevant aspects of sustainability and the possible impacts of changes due to decarbonization of industry in the field of concrete repair.

The book provides a guide for those responsible for buildings and other reinforced concrete structures as they are engaged to design, construct, maintain and repair them. It reviews the present state of knowledge of corrosion of steel in concrete from the theory through to site investigations and then remedying the problem. There is also a section on building more corrosion resistant structures.

The aim of the book is to educate and guide engineers, surveyors and owners of structures so that they will have a clear idea of the problem and will know where to go to start finding solutions. No single book can be a complete course in any subject and there is no substitute for personal 'hands on' experience. There are many experienced engineers and corrosion experts who deal with these problems on a day-to-day basis. The reader is recommended to seek expert advice when dealing with subjects that are new to him. However, the aim of being an 'informed client' is to be recommended and this book should help in that respect. It is also essential that students of civil engineering, building sciences and architecture understand the problems they will face as they start to practice their skills. We need more engineers, materials scientists, building surveyors and architects who are trained to look at wider durability issues in these days of sustainable development and conservation of resources.

It is impossible in one book to be totally comprehensive. Consequently some subjects such as cement chemistry and concrete admixtures have not been covered in much detail. They are better covered in other specialist texts. This book concentrates on the corrosion of steel in its cementitious environment and how it is dealt with in methods that are sometimes unique to the corrosion engineering field rather than the civil or building engineering field. I have been gratified by the reception of the previous editions by fellow engineers, scientists and others in the field and hope that this edition continues to be useful to the industry.

Illustration credits

A number of figures are reproduced in this book by kind permission of the copyright holders or authors as appropriate. In some cases acknowledgement is made in the figure caption; in the other cases the Publisher would like to gratefully acknowledge the permissions granted as follows:

Permission to reproduce extracts of BS EN 206:2013+A2:2021 and BS EN 1504-9:2008 in Tables 8.3 and 10.1 is granted by BSI. British Standards can be obtained from BSI Customer Services, 389 Chiswick High Road, London W4 4AL. Tel: +44 (0)20 8996 9001. email: cservices@bsi-global.com

Acknowledgements

It is difficult to know where to start and end with acknowledgements. I started my career after finishing University at the Central Electricity Research Laboratory, working in one of the largest groups of corrosion scientists in the UK and Europe. Sadly that group was dispersed with the privatisation of the Central Electricity Generating Board. I learned the basics of corrosion from Mike Manning, Jonathan Forrest and Ed Metcalfe and many others at CERL.

I then joined Taylor Woodrow and enjoyed carrying out some of the first trials of impressed current cathodic protection on reinforced concrete structures above ground in the UK, Hong Kong and Australia. I learned enormous amounts about deterioration of reinforced concrete structures, civil engineering and contracting from Roger Browne, Roger Blundell, Roger McAnoy, Phil Bamforth and a powerful team of engineers and scientists who were all at the forefront of concrete technology. I continue to have good relations and helpful exchanges with the Taywood team.

I was then privileged to be invited to work at the Strategic Highway Research Program (SHRP) in Washington DC. For three years we coaxed a $150 million research programme into life, and after five years we then shut it all down having spent the allotted budget. My thanks go to Damian Kulash who led the programme and was brave enough to take on a foreigner to oversee the Structures research and to Jim Murphy and Don Harriott who led the Concrete & Structures Group and taught me about highways. I must also thank all my colleagues at SHRP, the researchers who did the real research work and the advisory and expert committee members who gave their time and insights to help create some very valuable manuals and guidance for highway engineers.

The first draft of the first edition of this book was produced three years after returning to the UK as an independent consultant. I had the chance to work with some of the leading experts in the field of corrosion and deterioration of reinforced concrete. I picked their brains mercilessly and enjoyed working and relaxing with many such as Chris Wozencroft, Richard Edwards, Kevin Davies, Chris Atkins, Brian Wyatt, Nick Buenfeld,

Gareth Glass, Ken Boam, Mike Gower and Peter Johnson, the late David Whiting, Ken Clear, Jack Bennett, Carmen Andrade, Jesús Rodríguez and many others.

In the years since the previous edition I have been fortunate to continue to work with many knowledgeable and experienced engineers and specialists on the committees of the Association of Materials Protection and Performance AMPP, previously NACE, the Corrosion Prevention Association, Concrete Society, Institute of Corrosion, the Transportation Research Board and CEN. I must specifically acknowledge Andrew Trafford of Aperio who kindly supplied a lot of useful information as well as photographs about radar, radiography, pulse velocity, ultrasonic and impact-echo techniques. My thanks also go to my publisher, and to Nick Clarke whose helpful suggestions strengthened the first draft of the book considerably.

Finally, eternal thanks to my wife, Veronica and my parents, Olive and Philip. Everything one says about the support of the family sounds glib and clichéd, but it is, none the less, true. Without my parents' support I would not have got where I am today. Veronica's support has been invaluable in quietly encouraging me to get on with the job, make sure that I am focussed on the task in hand and not involved in pie in the sky ideas. She has supported my activities while getting on with her own activities.

My apologies for naming such a small number of the many people who I continue to have the pleasure and privilege of working with. My thanks to all of them and their willingness to share their knowledge with me and many others. For this third edition I should additionally thank Chris Atkins and Ali Sharifi who responded very kindly to my inquiries about where various technologies are going in the industry. Finally, it is sad to acknowledge that a number of those mentioned are either fully retired or no longer with us. But we remember their contributions and their friendship.

It is better to wear out than to rust out
Richard Cumberland, English Divine (1631–1718)

Paint, graffiti, and everything else that man can do cannot win against rust. Rust trumps all.
Laura Moncur commenting on a photograph by Ward Jenkins

Chapter 1

Introduction

Reinforced concrete is a versatile, economical and successful construction material. It can be moulded to a variety of shapes and finishes. In most cases it is durable and strong, performing well throughout its service life. However, in some cases it does not perform adequately due to poor design, poor construction, inadequate materials selection, a more severe environment than anticipated or a combination of these factors.

The corrosion of the reinforcing steel in concrete is a major problem facing civil engineers today as they maintain an ageing infrastructure. Potentially it is a very large market for those who develop the expertise to deal with the problem. It is also a major headache for those who are responsible for dealing with structures suffering from it.

Worldwide, more concrete is used than any other man-made material. Global cement production is expected to increase from 3.27 billion tonnes in 2010 to 4.83 billion tonnes in 2030 according to figures published by Statistica.com in February 2021. Correctly selected intervention that repairs a concrete structure cost effectively helps in conserving resources and reducing pollution from cement and concrete manufacturing.

One cost estimate was that there was $150 billion worth of corrosion damage on US interstate highway bridges due to deicing and sea salt-induced corrosion. In a Transportation Research Board report on the costs of deicing (Transportation Research Board, 1991) the annual cost of bridge deck repairs were estimated to be $50–200 million, with substructures and other components requiring $100 million a year and a further $50–150 million a year on multi-storey car parks.

In a cost of corrosion study (Koch *et al.*, 2002), the annual direct cost of corrosion on US highway bridges was estimated at $8.3 billion overall, with $4.0 billion of that on the capital cost and maintenance of reinforced concrete highway bridge decks and substructures. Indirect costs due to traffic delays were calculated to be more than 10 times the direct costs.

In the United Kingdom, the Department of Transport's (DoT) estimate of salt-induced corrosion damage was a total of £616.5 million on motorway and trunk road bridges in England and Wales alone (Wallbank, 1989).

DOI: 10.1201/9781003223016-1

These bridges represent about 10% of the total bridge inventory in the country. The eventual cost may therefore be 10 times the DoT estimate. The statistics for Europe, the Asian Pacific and Australia are similar.

In the Middle East the severe conditions of warm marine climate, especially in the more heavily populated areas, with saline ground waters, increases all corrosion problems. This is made worse by the difficulty of curing concrete that has led to very short lifetimes for reinforced concrete structures (Rasheeduzzafar *et al.*, 1992).

In many countries with rapidly developing infrastructure, economies or poor supervision and quality control procedures in construction have led to poor quality concrete and low concrete cover to the steel leading to carbonation problems.

Corrosion became a fact of life as soon as man started digging ores out of the ground, smelting and refining them to produce the metals that we use so widely in the manufacturing and construction industries. When man has finished refining the steel and other metals that we use, nature sets about reversing the process. The refined metal will react with the non-metallic environment to form oxides, sulphates, sulphides, chlorides and so on, which no longer have the required chemical and physical properties of the consumed metal. Billions of dollars are spent every year on protecting, repairing and replacing corrosion damage. Occasionally lives are lost when steel pipes, pressure vessels or structural elements on bridges fail. But corrosion is a slow process and is usually easy to detect before catastrophic failure, and the consequences of corrosion are usually economic rather than death or injury.

The corrosion of steel in concrete used to be considered to cause engineering and economic problems. However, there have been cases of large pieces of concrete falling from bridges in North America, with the loss of a motorist's life in New York City. The failure of post-tensioning tendons in a bridge caused a bridge collapse with the loss of a tanker driver's life in Europe and another failure of a pedestrian bridge in the United States. Car park collapses (Simon, 2004) and other major failures have been recorded in the United States, Canada and the United Kingdom as well as in Europe and the Far East (not always directly attributed to reinforcement corrosion).

The economic loss and damage caused by the corrosion of steel in concrete makes it arguably the largest single infrastructure problem facing industrialized countries. Our bridges, public utilities, chemical plants and buildings are ageing. Some can be replaced, others would cause great cost and inconvenience if they were taken out of commission. With major political arguments about how many more bridges, power plants and other structures we can build, it becomes crucial that the existing structures perform to their design lives and limits and are maintained effectively.

One of the biggest causes of corrosion of steel in concrete is the use of deicing salts on our highways. In the United States approximately 10 million tonnes of salt are applied per year to highways. In the United Kingdom 1–2 million tonnes per year are applied (but to a proportionately far smaller road network). In continental Europe the application rates are comparable, except where it is too cold for salt to be effective, or the population density too low to make salting worthwhile (e.g. in northern Scandinavia). However, research in the United States has shown that the use of deicing salt is still more economic than the alternative, more expensive, less effective deicers (Transportation Research Board, 1991).

In the early years of the 20th century corrosion of steel in concrete was attributed to stray current corrosion from electric powered vehicles. It was only in the 1950s in the United States that it was finally accepted that there were corrosion problems on bridges far from power lines but in areas where sea salt or deicing salts were prevalent. During the 1960s attempts were made to quantify the problem and in the 1970s the first cathodic protection systems were installed to control the corrosion of reinforcing steel in concrete. In Europe and the Middle East sea water for mixing concrete and the addition of calcium chloride set accelerators were acceptable until the 1960s and still used until the 1970s in a mistaken belief that most of the chlorides would be bound up in the cement and would not cause corrosion. This was found to be an expensive error over the next 20 years, particularly in the Middle East, where the low availability of potable water led to the frequent use of sea water for concrete mixing.

While corrosion of steel in concrete is a major cause of deterioration, it is not the only one. Out in the real world we must not become blinkered to other problems like alkali-silica reactivity, freeze thaw damage and the structural implications of the damage done and of repairs. In this setting, however, we will concentrate on the corrosion issue although there will be passing references to other problems where relevant.

There is some discussion in the book of the results of the Strategic Highway Research Program (SHRP). The author was a member of the SHRP staff for the first three years and a consultant for the last three years of SHRP. About $10 million worth of research was carried out over six years (1987–1993) on corrosion problems on existing reinforced concrete bridges. A complete list of SHRP reports on corrosion of steel in concrete is now available on the Transportation Research Board website (www.trb.org). SHRP was probably the largest single effort to address the problem of corrosion of steel in concrete and provide practical solutions for engineers to use.

In the United Kingdom the 'Concrete In the Oceans' research programme in the 1970s did much to improve our understanding of the fundamentals of corrosion of steel in concrete, particularly in marine conditions (Wilkins and Lawrence, 1980). Government laboratories around the world have worked on the problems since the 1960s when they started to manifest themselves.

The text draws on corrosion problems and experience from around the world where possible, attempting to achieve a balanced view of different approaches in different countries, particularly comparing the European and North American approaches which are sometimes quite different. Inevitably it concentrates on the author's primary experience in the United Kingdom, North America and, to a lesser extent, Europe, Australasia, the Middle and Far East and Africa.

Chapters 2 and 3 explore the basics of corrosion of steel in concrete. The author has attempted to be thorough but also to start from the position of the generalist, with a minimal memory of chemistry.

The first requirement when addressing a deterioration problem is to quantify it. Chapter 4 discusses condition evaluation and the testing procedures and techniques we can use to assess the causes and extent of the corrosion damage on a structure. Chapter 5 covers permanent corrosion monitoring, a subject that has grown considerably since the first edition of the book.

Chapters 6 and 7 are then concerned with repair and rehabilitation options, first the conventional physical intervention of concrete repair, patching, overlaying and coatings. The electrochemical techniques of cathodic protection, chloride extraction and realkalization are dealt with in Chapter 7.

Chapter 8 deals with how to select a rehabilitation method. It discusses the different techniques for repair covered in the previous chapters and the cost differentials. Chapter 9 reviews models for determining deterioration rates and service lives.

The penultimate chapter turns away from the issues of dealing with the problems on existing structures and reviews the options for new construction to ensure that the next generation of structures do not show the problems associated with some of the present generation. Many materials and methods of improving durability are on the market, but which ones are cost effective?

The final chapter discusses future developments. As materials science, computer technology, electronics and other disciplines advance, new methods for assessing corrosion are becoming available. The author speculates about what will be on offer and perhaps more importantly what we need to efficiently assess and repair corrosion damage on our fixed infrastructure in the future.

References

Frost, A. (2005). 'The Material World'. *New Civil Engineer*, Concrete Engineering Supplement, p 24.

Koch, G.H., Brogers, P.H., Thompson, N., Virmani, Y.P. and Payer, J.H. (2002). Corrosion Cost and Preventive Strategies in the United States. FHWA Report FHWA-RD-01–156. Federal Highway Administration, Washington, DC.

Rasheeduzzafar, H.S., Dakhil, F.H., Bader, M.A. and Khan, M.N. (1992). 'Performance of Corrosion Resisting Steel in Chloride-Bearing Concrete'. *ACI Materials Journal*, 89(5): 439–448.

Simon, P. (2004). Improved Current Distribution due to a Unique Anode Mesh Placement in a Steel Reinforced Concrete Parking Garage Slab CP System. NACE Corrosion 2004. Paper No. 04345.

Transportation Research Board (1991). Highway Deicing: Comparing Salt and Calcium Magnesium Acetate. Special Report 235. National Research Council, Washington, DC.

Wallbank, E.J. (1989). The Performance of Concrete in Bridges: A Survey of 200 Highway Bridges. HMSO, London.

Wilkins, N.J.M. and Lawrence, P.F. (1980). Concrete in the Oceans: Fundamental Mechanisms of Corrosion of Steel Reinforcements in Concrete immersed in Sea Water. Technical Report 6. ClRIA/UEG Cement and Concrete Association, Slough, UK.

Chapter 2

Corrosion of steel in concrete

This chapter discusses the basics of corrosion and how it applies to steel in concrete. To start with a basic question: Why does steel corrode in concrete? A more sensible question is why does steel not corrode in concrete? We know from experience that ordinary carbon steel reinforcing steel bars corrode (rust) when air and water are present. As concrete is porous and contains moisture, why doesn't steel in concrete usually corrode?

The answer is that concrete is alkaline. Alkalinity is the opposite of acidity. Generally, metals corrode in acids and they can be protected from corrosion by alkalis. This is the case for steel in concrete.

When we say that concrete is alkaline we mean that the concrete contains microscopic pores which contain high concentrations of soluble calcium, sodium and potassium oxides. These form hydroxides, which are very alkaline, when water is present. This creates a very alkaline condition of pH 12–13. The composition of the pore water and the movement of ions and gases through the pores is very important when analyzing the susceptibility of reinforced concrete structures to corrosion. This will be discussed in Chapter 3.

The alkaline condition leads to a 'passive' layer forming on the steel surface. The passive layer is a dense, impenetrable film, which, if fully established and maintained, prevents further corrosion of the steel. The layer formed on steel in concrete is probably part metal oxide/hydroxide and part mineral from the cement. A true passive layer is a very dense, thin layer of oxide that leads to a very slow rate of oxidation (corrosion). There is some discussion as to whether or not the layer on the steel is a true passive layer as it seems to be thick compared with other passive layers and it consists of more than just metal oxides, but as it behaves like a passive layer it is generally referred to as such.

Corrosion scientists and engineers spend much of their time trying to find ways of stopping corrosion of steel by applying protective coatings. Other metals such as zinc, polymers such as acrylics or epoxies are used to stop corrosive conditions getting to steel surfaces. The passive layer is the corrosion engineer's dream coating as it forms itself and will maintain and

DOI: 10.1201/9781003223016-2

repair itself as long as the passivating (alkaline) environment is there to regenerate it if it is damaged. If the passivating environment can be maintained it is far better than any artificial coatings like galvanizing or fusion bonded epoxy that can be damaged or consumed, allowing corrosion to proceed in damaged areas.

However, the passivating environment is not always maintained. Two conditions can break down the passivating environment in concrete without attacking the concrete first. One is carbonation and the other is chloride attack. These will be discussed in Chapter 3. In the rest of this chapter we will discuss what happens when depassivation has occurred.

2.1 The corrosion process

Once the passive layer breaks down then areas of rust will start appearing on the steel surface. The chemical reactions are the same whether corrosion occurs by chloride attack or carbonation. When steel in concrete corrodes it dissolves in the pore water and gives up electrons:

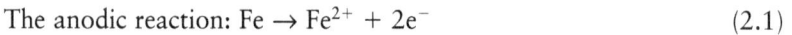

$$\text{The anodic reaction: } Fe \rightarrow Fe^{2+} + 2e^- \tag{2.1}$$

The two electrons ($2e^-$) created in the anodic reaction must be consumed elsewhere on the steel surface to preserve electrical neutrality. In other words we cannot have large amounts of electrical charge building up at one place on the steel. There must be another chemical reaction to consume the electrons. This is a reaction that consumes water and oxygen:

$$\text{The cathodic reaction: } 2e^- + H_2O + \tfrac{1}{2}O_2 \rightarrow 2OH^- \tag{2.2}$$

This is illustrated in Figure 2.1. You will notice that we are generating hydroxyl ions in the cathodic reaction. These ions increase the local alkalinity and therefore will strengthen the passive layer, warding off the effects of carbonation and chloride ions at the cathode. Note that water and oxygen are needed at the cathode for corrosion to occur.

The anodic and cathodic reactions ((2.1) and (2.2)) are only the first steps in the process of creating rust. However, this pair of reactions is critical to the understanding of corrosion and is widely quoted in any discussion on corrosion and corrosion prevention for steel in concrete. The reactions will be referred to often in this book.

If the iron were just to dissolve in the pore water (the ferrous ion Fe^{2+} in equation (2.1) is soluble) we would not see cracking and spalling of the concrete. Several more stages must occur for 'rust' to form. This can be expressed in several ways and one is shown here where ferrous hydroxide becomes ferric hydroxide and then hydrated ferric oxide or rust:

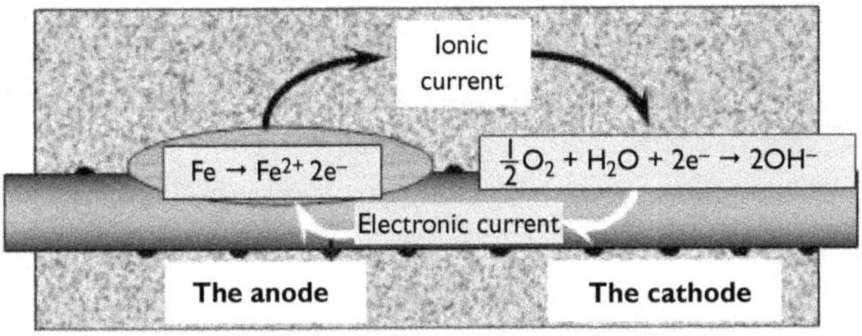

$Fe^{2+} + 2OH^- \rightarrow Fe(OH)_2$ Ferrous hydroxide
$4Fe(OH)_2 + O_2 + 2H_2O \rightarrow 4Fe(OH)_3$ Ferric hydroxide
$2Fe(OH)_3 \rightarrow Fe_2O_3 \cdot H_2O + 2H_2O$ Hydrated ferric oxide (rust)

Figure 2.1 The anodic, cathodic and oxidation and hydration reactions for corroding steel.

$$Fe^{2+} + 2OH^- \rightarrow Fe(OH)_2 \qquad (2.3)$$
$$\text{Ferrous hydroxide}$$

$$4Fe(OH)_2 + O_2 + 2H_2O \rightarrow 4Fe(OH)_3 \qquad (2.4)$$
$$\text{Ferric hydroxide}$$

$$2Fe(OH)_3 \rightarrow Fe_2O_3{\cdot}H_2O + 2H_2O \qquad (2.5)$$
$$\text{Hydrated ferric oxide (rust)}$$

The full corrosion process is illustrated in Figure 2.1. Unhydrated ferric oxide Fe_2O_3 has a volume of about twice that of the steel it replaces when fully dense. When it becomes hydrated it swells even more and becomes porous. This means that the volume increase at the steel/concrete interface is six to ten times as indicated in Figure 2.2. This leads to the cracking and spalling that we observe as the usual consequence of corrosion of steel in concrete and the red/brown brittle, flaky rust we see on the bar and the rust stains we see at cracks in the concrete.

Several factors in the explanation given in this section are important and will be used later to explain how we measure and stop corrosion. The electrical current flow, and the generation and consumption of electrons in the anode and cathode reactions, are used in half-cell potential measurements and cathodic protection. The formation of protective, alkaline hydroxyl ions is used in cathodic protection, electrochemical chloride removal and realkalization. The fact that the cathodic and anodic reactions must balance each other for corrosion to proceed is used in epoxy coating protection of rebars. All this will be discussed later.

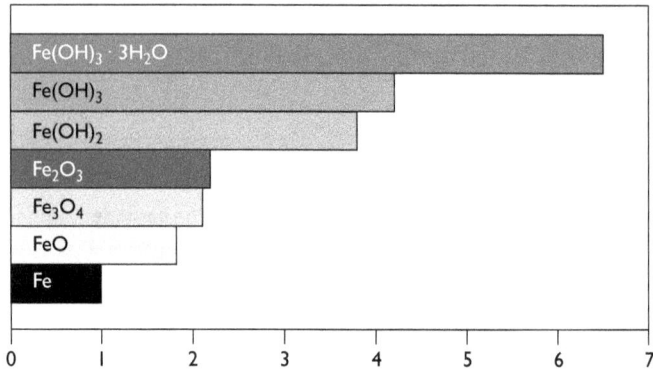

Figure 2.2 Relative volume of iron and its oxides.

2.2   Black rust

There is an alternative to the formation of 'normal' red rust described in reactions (2.3) to (2.5) earlier. If the anode and cathode are well separated (by several hundred millimetres) and the anode is starved of oxygen (say by being underwater) the iron as Fe^{2+} will stay in solution. This means that there will be no expansive forces as described earlier to crack the concrete so corrosion may not be detected.

This type of corrosion (known as 'black' or 'green' rust due to the colour of the liquid when first exposed to air after breakout) is found under damaged waterproof membranes and in some underwater or other water saturated conditions. It is potentially dangerous as there is no indication of corrosion by cracking and spalling of the concrete and the reinforcing steel may be severely weakened before corrosion is detected. Rebars may be hollowed out in such deoxygenated conditions particularly under membranes or when water is permanently ponded on the surface.

Examples of rebars attacked in this way are shown in Figure 2.3. These bars were taken from underneath damaged waterproof membranes. Rust staining on the concrete surface may be indicative of this type of attack, but obviously if water is getting under a membrane and excluding oxygen it is unlikely that the iron in solution will get to the concrete surface where it will then precipitate out to form rust stains.

2.3   Pits, stray current and bacterial corrosion

2.3.1   Pit formation

Corrosion of steel in concrete generally starts with the formation of pits. These increase in number, expand and join up leading to the generalized

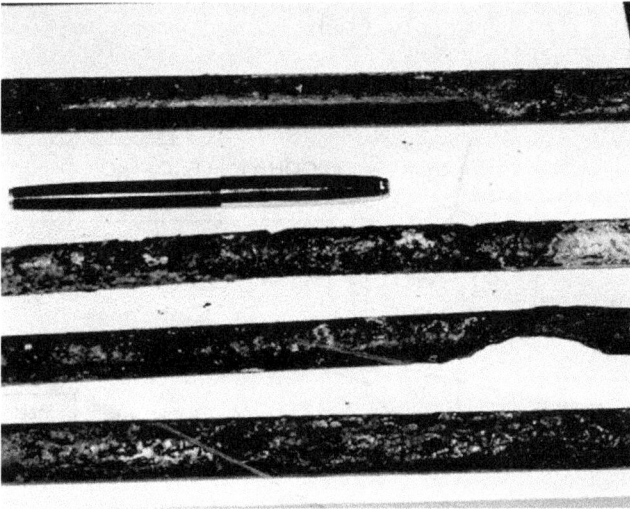

Figure 2.3 Reinforcing bars taken from under the end of a waterproofing membrane. They have been subjected to low oxygen conditions and therefore one of them shows severe local wasting. There was no expansive oxide growth.

corrosion usually seen on reinforcing bars exposed to carbonation or chlorides. The formation of pits is illustrated in Figure 2.4.

The chemistry of pitting is quite complex and is explained in most chemistry text books. However, the principle is fairly simple, especially where chlorides are present. At some suitable site on the steel surface (often thought to be a void in the cement paste or a sulphide inclusion in the steel) the passive layer is more vulnerable to attack and an electrochemical potential difference exists that attracts chloride ions. Corrosion is initiated and acids are formed, hydrogen sulphide from the MnS inclusion and HCl from the chloride if they are present. Iron dissolves (equation (2.1)), the iron in solution reacts with water:

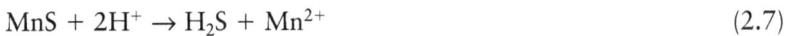

$$Fe^{2+} + H_2O \rightarrow FeOH^+ + H^+ \tag{2.6}$$

$$MnS + 2H^+ \rightarrow H_2S + Mn^{2+} \tag{2.7}$$

A pit forms, rust may form over the pit, concentrating the acid (H^+), excluding oxygen so that the iron stays in solution preventing the formation of a protective oxide layer and accelerating corrosion.

We will return to the subject of pitting corrosion later. It is related to the problems of coated reinforcement and to the 'black rust' phenomenon discussed earlier.

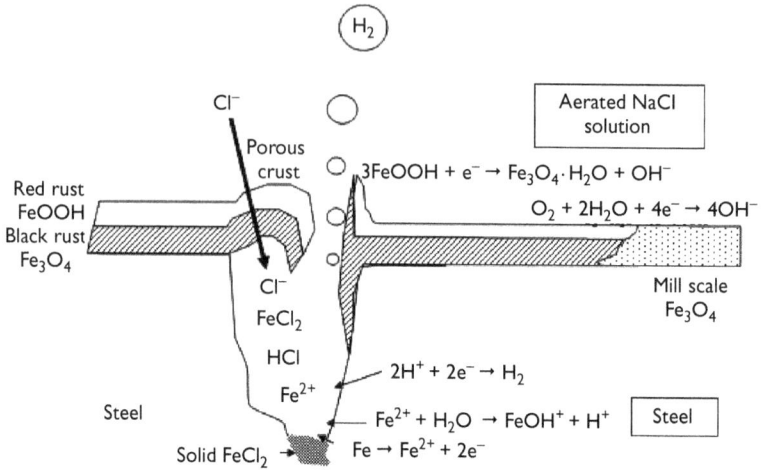

Figure 2.4 The classical corrosion model of pitting attack adapted from Wranglen G. *Corrosion Sci*, 1974, 14: 331.

2.3.2 Bacterial corrosion

Another complication comes from bacterial corrosion. There are bacteria in soil (thiobaccilli) that convert sulphur and sulphides to sulphuric acid. There are other species (ferrobaccili) that attack the sulphides in steel (FeS). This is often associated with a smell of hydrogen sulphide (rotten eggs) and smooth pitting with a black corrosion product when rebars are exposed having been in water saturated conditions. In anaerobic (oxygen starved) conditions such bacteria can contribute to the pitting corrosion discussed earlier.

The alkaline conditions in concrete generally keep bacteria at bay. However, they have been seen on steel in damaged concrete where the steel surface is no longer alkaline and oxygen is excluded by saturation.

Bacterial corrosion of steel should not be confused with bacterial attack of concrete. This can occur in sewer systems. A summary of the issues of microbial induced 'corrosion' of concrete is given in Lines *et al.* (2021).

2.3.3 Stray-current-induced corrosion

As stated in the introduction, stray currents were originally blamed for corrosion in concrete in the United States until the problem of chloride attack was identified in the 1950s. The main cause of stray-current-induced corrosion was the direct current from DC traction systems on trams (streetcars), when current flowed through buried or embedded steel work. In finding the lowest resistance path to ground, current will jump from one metal conductor to another via an ionic medium when the metals are not in contact. This might be from one reinforcement cage to another via the concrete pore water. One end of the reinforcement cage will become negative (a cathode), and will not corrode. The other end will become positive and will actively corrode. The problems of stray currents in cathodic protection systems are discussed in Section 7.8.2 where the effect of having electrically discontinuous steel in the structure is considered.

Today, stray current corrosion in bridges, buildings and above ground structures is a specialized problem dealt with by engineers designing and maintaining light rail systems. Below ground the problem also occurs on cathodic protection systems for pipelines and other structures. Cathodic protection systems on buried pipelines can interact with adjacent or crossing facilities. There must be an ionic path between the source of current and the metal at risk which must also be on the path of lowest resistance. For this reason stray-current-induced corrosion is most unlikely on above ground structures when the DC source is below ground.

There is some discussion in the literature about AC induced stray currents. These are generally far lower than DC effects although recent literature suggests that in some cases it can be significant (Gummow *et al.*, 1998). One case where AC can be significant is on a structure with cathodic protection installed. In such conditions, the AC can shift the steel potential to the actively corroding regions of the electrochemical spectrum causing corrosion. NACE/AMPP have published a report and a standard practice on the subject of stray-current-induced corrosion of steel in concrete (NACE 2019 a and b). These show that the passive layer is good at protecting against stray currents, but the presence of chloride will accentuate stray-current-induced corrosion.

2.3.4 Local vs. general corrosion (macrocells vs. microcells)

Corrosion is often local, with a few centimetres of corrosion and then up to a metre of clean passive bar, particularly for chloride-induced corrosion. This indicates the separation of the anodic reaction (2.1) and the cathodic reaction (2.2) to form a 'macrocell'. Chloride-induced corrosion gives rise to particularly well defined macrocells. This is partly due to the mechanism of chloride attack, with pit formation and with small concentrated anodes being 'fed' by large cathodes. It is also because chloride attack is usually associated with high levels of moisture giving low electrical resistance in the concrete and easy transport of ions so the anodes and cathodes can separate easily.

In North America, this is used as a way of measuring corrosion by measuring 'macrocell currents'. This is discussed later in Section 4.12.6.

2.4   Electrochemistry, cells and half cells

The first two reactions we discussed in this chapter were the anodic and cathodic reactions for steel in concrete. The terms 'anode' and 'cathode' come from electrochemistry which is the study of the chemistry of electrical cells. Figure 2.5 is a basic Daniell cell which is used at high school to illustrate how chemical reactions produce electricity. The cell is composed of two 'half cells', copper in copper sulphate and zinc in zinc sulphate. The total voltage of the cell is determined by the metals used and by the nature and composition of the solutions. What is happening is that in each half cell the metal is dissolving and ions are precipitating, that is,

$$M \longleftrightarrow M^{n+} + ne^-$$

Copper is more resistant to this reaction than zinc so when we connect the two solutions by a semi permeable membrane and connect the two metals with a wire, the zinc goes into solution, and the copper sulphate plates out as pure copper on the copper electrode.

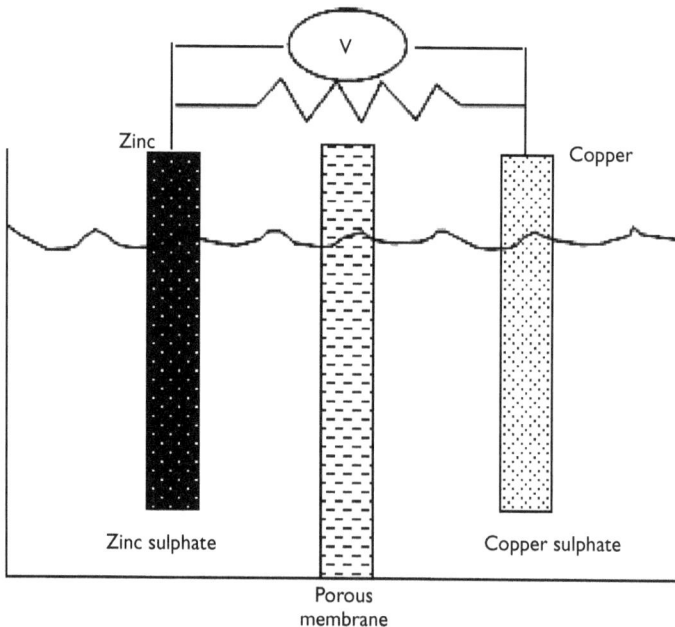

Figure 2.5 The Daniell cell. Cell voltage $= 0.34 - (-0.76) = 1.10\,V$.

Table 2.1 Standard reference electrode potentials

$Zn \rightarrow Zn^{2+} + 2e^-$	$-0.76\,V$
$Fe \rightarrow Fe^{2+} + 2e^-$	$-0.44\,V$
$Cu \rightarrow Cu^{2+} + 2e^-$	$+0.34\,V$

The voltage of any half cell can be recorded against a standard hydrogen electrode (half cell or reference electrode). Table 2.1 gives the standard reference electrode or half cell potentials that are of interest to us as we evaluate corrosion problems.

Reference electrode potentials are a function of solution concentration as well as the type of metal and the solution. A more concentrated solution is (generally) more corrosive than a dilute one so a current will flow in a cell made up of a single metal in two different concentrations of the same solution. We can consider the corrosion of steel in concrete as a concentration cell.

We can measure the corrosion risk in the cell by introducing an external reference electrode. This is most easily illustrated by a copper/saturated copper sulphate reference electrode moved along the surface of the concrete with a rebar in the concrete which has anodic (corroding) areas and cathodic (passive) areas.

As we see in Figure 2.6, by placing a reference electrode on the concrete surface and connecting it via a voltmeter to the steel, we have a similar circuit to our Daniell cell (Figure 2.5). The electrical potential difference will be a function of the iron in its pore water environment. If we move the cell along the steel we will see different potentials because the iron is in different environments. At the anode it can easily go into solution like the zinc in our Daniell cell. At the cathode, the passive layer is still strong and being strengthened further by the cathodic reaction, so the steel resists dissolution.

As a result we see higher potentials (voltages) on our voltmeter in anodic, corroding areas and lower voltages in the cathodic, passive areas. The use and interpretation of reference electrode potential measurements is discussed further in Section 4.8.

We must be careful in our use of electrochemical theory to explain what is going on in a corrosion cell. Electrochemical theory generally applies to equilibrium conditions and well defined solutions. Corrosion is not an equilibrium, but a dynamic situation. Therefore using the theory and equations of electrochemistry is an approximation and can lead to errors if the model is stretched too far.

2.5 Conclusions

This chapter has discussed the mechanism of what happens at the steel surface. The chemical reactions, formation of oxides, pitting, stray currents,

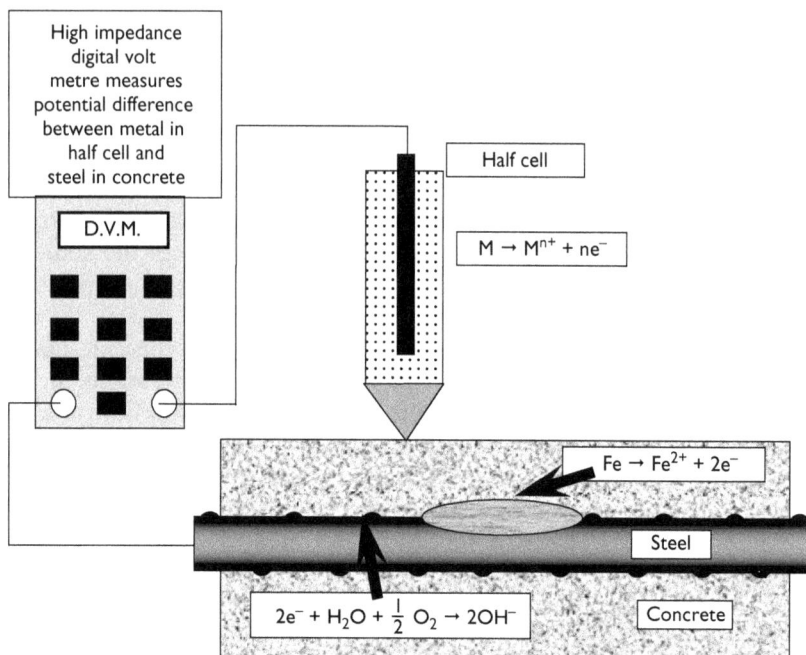

Figure 2.6 Schematic of a reference electrode (half cell) potential measurement of steel in concrete.

bacterial corrosion, anodes, cathodes and reference electrode potentials (half cells) have been reviewed. A more detailed account of the electrochemistry of corrosion and corrosion of steel in concrete is given in Appendix B. Chapter 3 will discuss the processes that lead to the corrosion and the consequences in terms of damage to structures. We will then move on to the measurement of the problem and how to deal with it.

References

Gummow, R.A., Wakelin, R.G. and Segall, S.M. (1998). AC Corrosion – A New Challenge to Pipeline Integrity. NACE Corrosion 98. Paper No. 566.

Lines, S.J., Rothstead, D.A., Rollins, B. and Alt C. (2021). 'Microbially Induced Corrosion of Concrete'. *Concrete International*, 43(5): 28–31.

NACE (2019a). SP 21427 2019 Detection and Mitigation of Stray Current Corrosion of Reinforced and Prestressed Concrete Structures.

NACE (2019b). Publication 01110-2019 Stray-Current-Induced Corrosion in Reinforced and Prestressed Concrete Structures.

Chapter 3

Causes and mechanisms of corrosion in concrete

There are two main causes of corrosion of steel in concrete. This chapter will discuss how chloride attack and carbonation lead to corrosion and how the corrosion then proceeds. There will also be discussion of the variations that can be found when carrying out investigations in the field.

The main causes are chloride attack and carbonation. These two mechanisms are unusual in that they do not attack the integrity of the concrete. Instead, aggressive chemical species pass through the pores and attack the steel. Other acids and aggressive ions such as sulphate destroy the integrity of the concrete before the steel is affected. Most forms of chemical attack are therefore concrete problems before they are corrosion problems. Carbon dioxide is very unusual in penetrating the concrete without significantly damaging it. Accounts of (for instance) acid rain causing corrosion of steel embedded in concrete are unsubstantiated as far as this author is aware. While the carbonation process affects the microstructure of the concrete it is not generally deleterious.

3.1 Carbonation

Carbonation is the result of the interaction of carbon dioxide gas in the atmosphere with the alkaline hydroxides in the concrete. Like many other gases carbon dioxide dissolves in water to form an acid. Unlike most other acids the carbonic acid does not attack the cement paste, but just neutralizes the alkalies in the pore water, mainly forming calcium carbonate that lines the pores:

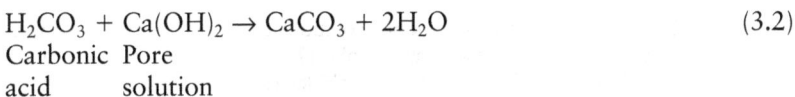

$$CO_2 + H_2O \rightarrow H_2CO_3 \qquad (3.1)$$
Gas   Water   Carbonic acid

$$H_2CO_3 + Ca(OH)_2 \rightarrow CaCO_3 + 2H_2O \qquad (3.2)$$
Carbonic Pore
acid   solution

DOI: 10.1201/9781003223016-3

There is a lot more calcium hydroxide in the concrete pores that can dissolve in the pore water. This helps maintain the pH at its usual level of 12–13 as the carbonation reaction occurs. However, as carbon dioxide proceeds to react with the calcium (and other) hydroxides in solution, eventually all the calcium hydroxide reacts, precipitating the calcium carbonate and allowing the pH to fall to a level where steel will corrode. This is illustrated in Figure 3.1(a) and (b) which show the reduction in the corrosion rate of steel as the pH increases and the pH changes across the carbonation front.

Carbonation damage occurs most rapidly when there is little concrete cover of the reinforcing steel. It can also occur when the cover is high but the pore structure is open, pores are well connected together and allow rapid CO_2 ingress and when alkaline reserves in the pores are low. This occurs when there is a low cement content, high water cement ratio and poor curing of the concrete.

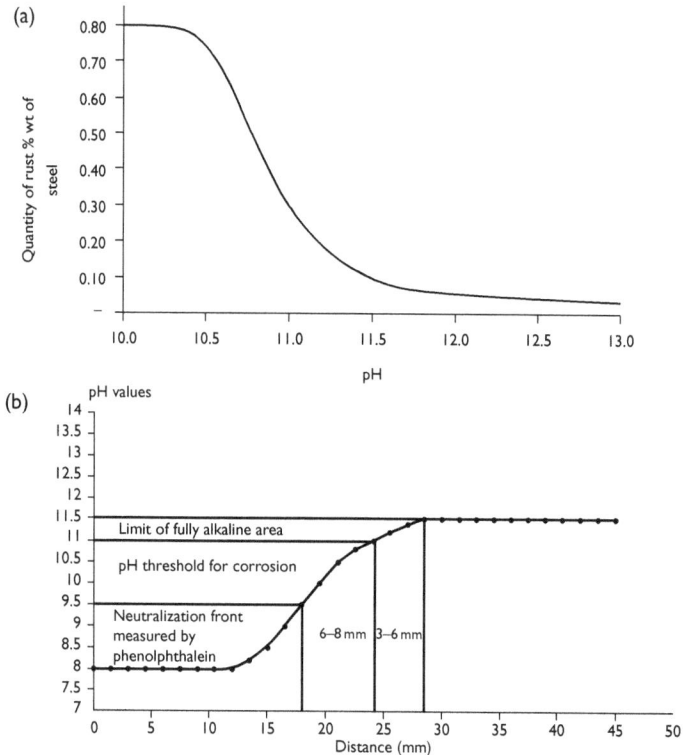

Figure 3.1 (a) Corrosion of steel in aqueous solutions as a function of pH, showing the effect of the passive layer above pH 12.5. Data from Shalon and Raphael *J ACI*, 1959, (6): 1251–1268. (b) The pH levels for carbonation, corrosion and phenolphthalein indicator.

A carbonation front proceeds into the concrete roughly following the laws of diffusion. These are most easily defined by the statement that the rate is proportional to the thickness:

$$\frac{dx}{dt} = \frac{D_o}{x} \qquad (3.3)$$

where x is distance, t is time and D_o is the diffusion constant.

The diffusion constant D_o is determined by the concrete quality. At the carbonation front there is a sharp drop in alkalinity from pH 11–13 down to less than pH 8. At that level the passive layer, which we saw in Chapter 2 was created by the alkalinity, is no longer sustained so corrosion proceeds by the general corrosion mechanism as described in the Chapter 2.

Many factors influence the ability of reinforced concrete to resist carbonation-induced corrosion. As the carbonation rate is a function of thickness, good cover is essential to resist carbonation. As the process is one of neutralizing the alkalinity of the concrete, good reserves of alkali are needed, that is, a high cement content. The diffusion process is made easier if the concrete has an open pore structure. On the macroscopic scale this means that there should be good compaction. On a microscopic scale well cured concrete has small pores and lower connectivity of pores, therefore the CO_2 has a harder job moving through the concrete. Microsilica and other additives can block pores or reduce pores sizes. However, most cement replacement materials have low alkali contents so there is a trade off between reduced diffusivity and lower alkali reserves. This is an issue as we move away from well characterized blended cements as for instance, pulverized fuel ash from coal fired power stations becomes less available, to new materials where the alkali content is not well characterized and the effect on alkali reserves is not well researched.

Carbonation is common on old structures, badly built structures (particularly buildings) and reconstituted stone elements containing reinforcement (often having a low cement content and being very porous). Carbonation is rare on modern highway bridges and other civil engineering structures where water cement ratios are low, cement contents are high with good compaction and curing and enough cover to prevent the carbonation front advancing into the concrete to any significant extent. On these structures the chlorides usually get there first. Wet dry cycling will accelerate carbonation by allowing CO_2 in and then supplying the water for it to dissolve in. This gives problems in some countries where the cycling between wet and dry seasons seems to favour carbonation.

When a repairer talks of repairing corrosion due to 'low cover' he usually means that the concrete has carbonated around the steel leading to corrosion.

As the cover is low it was a quick process. If the concrete were of the highest quality carbonation may not have been possible and low cover might not have mattered.

Carbonation is easy to detect and measure. A pH indicator, usually phenolphthalein in a solution of water and alcohol, will detect the change in pH across a freshly exposed concrete face. Phenolphthalein changes from clear at low pH (carbonated zone) to pink at high pH (uncarbonated concrete). Measurements can be taken on concrete cores, fragments and down drilled holes. Care must be taken to prevent dust or water from contaminating the surface to be measured but the test, with the indicator sprayed onto the surface, is cheap and simple. Figure 3.1(b) shows a typical carbonation front in concrete.

3.1.1   Carbonation transport through concrete

As stated earlier, carbon dioxide diffuses through the concrete and the rate of movement of the carbonation front approximates to Fick's law of diffusion. This states that the rate of movement is proportional to the distance from the surface as in equation (3.3) earlier. However, as the carbonation process modifies the concrete pore structure as it proceeds this is only an approximation. Cracks, changes in concrete composition and moisture levels with depth will also lead to deviation from the perfect diffusion equation. Integration of equation (3.3) gives a square root law which can be used to approximate the movement of the carbonation front. The calculation of diffusion rates is discussed in more detail in Chapter 8.

Empirically, a number of equations have been used to link carbonation rates, concrete quality and environment. Table 3.1 summarizes some of those equations and shows the factors that have been included. Generally there is a $t^{1/2}$ dependence. As discussed earlier the other factors are exposure, water/cement ratio, strength and CaO content (both functions of cement and, that is, alkali content).

If we consider the basic equation:

$$d = At^{0.5}$$

In a study of a range of structures (Trend 2000 et al., 2001) the following was found:

- For 11 buildings 8–24-years-old, the constant A ranged from 1.2–6.7 with an average of 3 $y^{1/2}$/mm.
- For 7 car parks 14–41-years-old, the constant A ranged from 2.2–7.6 with an average 4.27 $y^{1/2}$/mm.
- For a 10-year-old jetty and a 90-year-old bridge A was 1.8 for the jetty and 1.6 for the bridge.

Table 3.1 A selection of carbonation depth equations

Equation	Parameters
$d = At^n$	d = carbonation depth t = time in years A = diffusion coefficient n = exponent (approximately $1/2$)

$d = A \cdot B \cdot C \cdot t^{0.5}$	A = 1.0 for external exposure
	    1.7 for internal exposure
	B = 0.07 to 1.0 depending on surface finish
	$C = R(wc - 0.25)/(0.3(1.15 + 3wc))^{1/2}$
	    for water cement ratio $(wc) \geqslant 0.6$
	$C = 0.37R(4.6wc - 1.7)$ for $wc < 0.6$
	R = coefficient of neutralization, a function of
	    mix design and additives

$d = A(B \cdot wc - C)t^{0.5}$	A is a function of curing B and C are a function of fly ash used
$d = 0.43(wc - 0.4)(12(t - 1))^{0.5} + 0.1$ $d = 0.53(wc - 0.3)(12t)^{0.5} + 0.2$ $d = (2.6(wc - 0.3)^2 + 0.16)t^{0.5}$ $d = (wc - 0.3)^2 + 0.07)t^{0.5}$ $d = 10.3e^{-0.123f28}$	28 day cured Uncured Sheltered Unsheltered (unsheltered at 3 years) $-$ (fX = strength at day X)
$d = 3.4e^{-0.34f28}$ $d = 680(f28 + 25)^{-1.5} - 0.6$ at 2 years $d = A + B/f28^{0.5} + c/(CaO - 46)^{0.5}$ $d = (0.508/f35^{0.5} - 0.047)(365t)^{0.5}$ $d = 0.846(10wc/(10f7)^{0.5} - 0.193$ $-0.076wc)(12t)^{0.5} - 0.95$	(sheltered) CaO is alkali content expressed as CaO
$d = A(T - t_i)T^{0.75}(C_1/C_2)^{0.5}$	t_i = induction time T = temperature in $^\circ K$ C_1 = CO_2 concentration C_2 = CO_2 bound by concrete

Source: Parrott, L.J. A Review of Carbonation in Reinforced Concrete. A review carried out by the Cement and Concrete Association under a BRE Contract Publ. BRE Garston, Watford, UK: 1987, Jul.

3.2   Chloride attack

3.2.1   *Sources of chlorides*

Chlorides can come from several sources. They can be cast into the concrete or they can diffuse in from the outside. Chlorides cast into concrete can be due to:

- deliberate addition of chloride set accelerators (calcium chloride $CaCl_2$ was widely used until the mid-1970s);
- use of seawater in the mix;
- contaminated aggregates.

Chlorides can diffuse into concrete due to:

- sea salt spray and direct seawater wetting;
- deicing salts;
- use of chemicals (structures used for salt storage, brine tanks, aquaria, etc.).

Much of our discussion will centre on the diffusion of chlorides into concrete as that is the major problem in most parts of the world either due to marine salt spray or use of deicing salts. However, the cast in chlorides must not be overlooked especially when they are part of the problem. This often happens in marine conditions where sea water contaminates the original concrete mix and then diffuses into the hardened concrete.

3.2.2   Chloride transport through concrete

Like carbonation, the rate of chloride ingress is often approximated to Fick's law of diffusion. There are further complications here. The initial mechanism appears to be suction, especially when the surface is dry, that is, capillary action. Salt water is rapidly absorbed by dry concrete. There is then some capillary movement of the salt laden water through the pores followed by 'true' diffusion. There are other opposing mechanisms that slow the chlorides down. These include chemical reaction to form chloroaluminates and absorption onto the pore surfaces. The detailed transport mechanisms of chloride ions into concrete are discussed in Kropp and Hilsdorf (1995).

The other problem with trying to predict the chloride penetration rate is defining the initial concentration as the chloride diffusion is a concentration gradient, not a front. In other words we can use the square root relationship for the carbonation front as the concrete either is or is not carbonated, but we cannot use it so easily for chlorides as there is no chloride 'front', but a concentration profile in the concrete. Typical chloride profiles are shown in Figure 3.2(a) and (b). Figure 3.2(a) shows a 'classical' diffusion curve. Figure 3.2(b) shows the more erratic profiles in a multi-storey car park. The higher chloride concentrations at the deepest increments may be close to the soffit of the slab where chloride is concentrating due to evaporation. The calculation of chloride diffusion rates is discussed more fully in Chapter 9.

The different mechanisms of chloride transport into concrete.

3.2.3   Chloride attack mechanism

In Chapter 2, we discussed the corrosion of steel in concrete and the effectiveness of the alkalinity in producing a passive layer of protective oxide on the steel surface which stops corrosion. In the previous section we observed that alkalinity is neutralized by carbonation. The depassivation mechanism

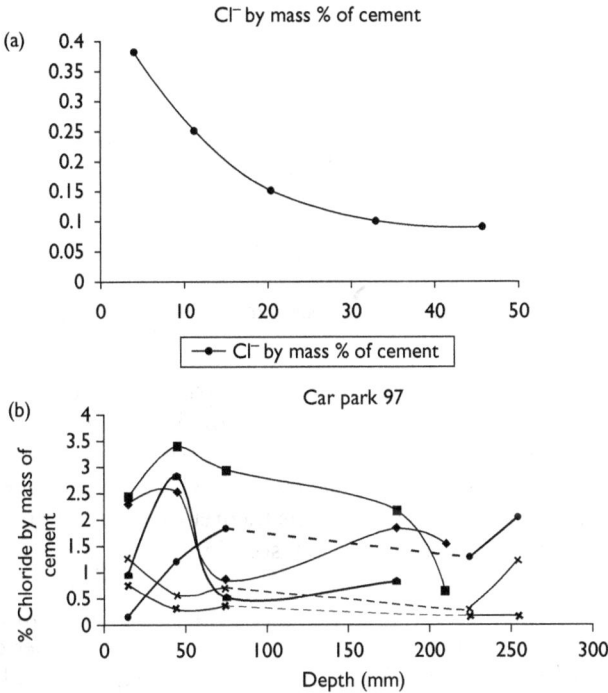

(a)

Cl⁻ by mass % of cement

0.4
0.35
0.3
0.25
0.2
0.15
0.1
0.05
0

0   10   20   30   40   50

— Cl⁻ by mass % of cement

(b)

Car park 97

% Chloride by mass of cement

4
3.5
3
2.5
2
1.5
1
0.5
0

0   50   100   150   200   250   300

Depth (mm)

Figure 3.2 (a) Chloride profile of a highway bridge from Broomfield, J. P. Field Survey of Cathodic Protection on North American Bridges. *Materials Performance*, 1992 Sep.; 31(9): 28–33. (b) Chloride profiles from a car park depth from Trend 2000; BRE; J Broomfield Consultants, and Risk Review Ltd. Structures Data Base – Project Data Base. Corrosion of Steel in Concrete – 2001 Dec; DTI DME 5.1. BRE.

for chloride attack is somewhat different. The chloride ion attacks the passive layer although in this case (unlike carbonation) there is no overall drop in pH. Chlorides act as catalysts to corrosion. They are not consumed in the process but help to break down the passive layer of oxide on the steel and allow the corrosion process to proceed quickly. This is illustrated in Figure 3.3, while the main pitting process is shown in Figure 2.4. The effective recycling of chloride ions makes chloride attack more difficult to remedy as chlorides are therefore harder to eliminate.

Obviously a few chloride ions in the pore water will not break down the passive layer, especially if it is effectively re-establishing itself when damaged as discussed in Chapter 2. There is a well known 'chloride threshold' for corrosion given in terms of the chloride/hydroxyl ratio. This was first investigated by Hausmann (1967).

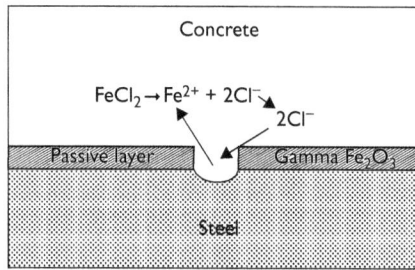

Figure 3.3 The breakdown of the passive layer and 'recycling' chlorides.

Hausmann (1976) used a simple Monte Carlo (random number) calculation to show that if chloride ions and hydroxyl ions are competing to fill a break in the passive layer, the chloride ion starts to break down the passive layer when the chloride concentration exceeds 0.6 of the hydroxyl concentration. Haussman compared his theoretical calculations with laboratory tests with calcium hydroxide solutions. The 0.6 Cl^-/OH ratio *approximates* to a concentration of 0.4% chloride by weight of cement if chlorides are cast into concrete and 0.2% if they diffuse in. In the United States a commonly quoted threshold is 1 pound of chloride per cubic yard of concrete. Although these figures are based on experimental evidence, the actual numbers are a function of practical observations of real structures.

All these thresholds are approximations because:

(a) concrete pH varies with the type of cement and the concrete mix. A tiny pH change is a massive change in OH^- concentration as pH is the logarithm of hydroxyl ion concentration and therefore the threshold moves radically with pH;

(b) chlorides can be bound chemically (by aluminates in the concrete) and physically (by absorption on the pore walls). This removes them (temporarily or permanently) from the corrosion reaction;

(c) in very dry concrete corrosion may not occur even at very high Cl^- concentration as the water is missing from the corrosion reaction as discussed in Chapter 2;

(d) in sealed, polymer impregnated or water saturated concrete, corrosion may not occur even at a very high Cl^- concentration as there is no oxygen present to fuel the corrosion reaction. This can also occur when there is total water saturation, but if some oxygen gets in then the pitting corrosion described in Chapter 2 can occur;

(e) macrocells develop on the steel surface which means that sampling in a cathodic area will show no corrosion despite high chloride levels.

Therefore corrosion can be observed at 0.2% chloride if the concrete quality is poor and none is seen above 1.0% or more if oxygen and water are excluded. If the concrete is very dry or totally saturated (as in (c) or (d)) then a change in conditions may lead to rapid corrosion.

In fact the threshold of corrosion should be considered to be a probabilistic function rather than a fixed number. Figure 3.4(a) shows the cumulative probability of corrosion vs. the chloride concentration on a UK bridge (Vassie, 1987). However, when differentiated (3.4b), it can be seen that there are 'thresholds' at 3%, 6% and 1.2% chloride by mass of cement. One possible reason is discussed in the next section.

3.2.4 Macrocell formation

As stated in Chapter 2, corrosion proceeds by the formation of anodes and cathodes (Figures 2.1 and 2.2). In the case of chloride attack they are often well separated with areas of rusting separated by areas of 'clean' steel. This is known as the macrocell phenomenon. Chloride-induced corrosion is particularly prone to macrocell formation as a high level of water is usually

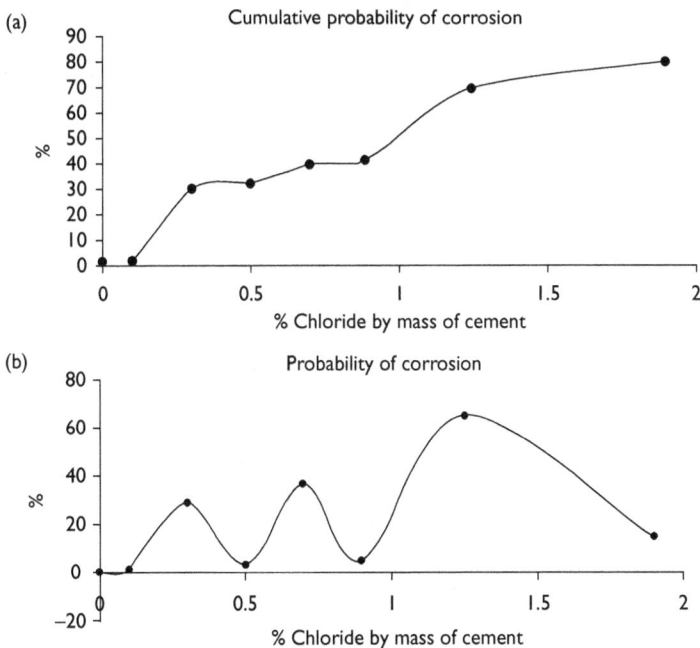

Figure 3.4 (a) The probability of finding corroding steel associated with a specified chloride concentration on a 20-year-old bridge deck (Vassie, 1987). (b) Differential of Figure 3.4(a) showing thresholds at 3%, 6% and 1.2% chloride by mass of cement.

Figure 3.5 Strongly differentiated anode and cathode regions or macrocell effects on a jetty substructure (new reinforcement has been placed to replace corroded stirrups prior to repair and the application of impressed current cathodic protection).

present to carry the chloride into the concrete and because chlorides in concrete are hygroscopic (i.e. they absorb and retain moisture). The presence of water in the pores increases the electrical conductivity of the concrete. The higher conductivity allows the separation of anode and cathode as the ions can move through the water filled (or water lined) pores.

The separation of corroded areas does not necessarily represent the distribution of chlorides along the rebar. It is the separation of anodic and cathodic reactions which is seen with large cathodic areas supporting small, concentrated anodic areas. The phenomenon is illustrated in Figure 3.5 from a jetty substructure in Jersey where anodes and cathode areas are strongly differentiated on the central bar.

The peaks seen in Figure 3.4(b) may correspond to the breaking down of pits into more generalized corrosion and the breaking down of macrocells into uniform corrosion.

For carbonation the concrete is generally dryer (otherwise the CO_2 does not penetrate far). Corrosion therefore tends to be on a 'microcell' level with apparently continuous corrosion along the reinforcing steel for carbonated concrete.

3.3   Corrosion damage

In most industries corrosion is a concern because of wastage of metal leading to structural damage. This can be collapse, perforation of containers and pipes, etc. However, most problems with corrosion of steel in concrete are not due to loss of steel but the growth of the oxide. This leads to cracking and spalling of the concrete cover.

Structural collapses of reinforced concrete structures due to corrosion are rare. The author knows of two multi-storey parking structures in North America which have collapsed due to deicing salt-induced corrosion (Simon, 2004). A prestressed concrete bridge collapsed in Wales due to deicing salt attack on the strands hidden from investigation (Woodward and Williams, 1988), and another one in Europe. More recently, there has been a condominium collapse in Florida and an unusual bridge design in Italy where corrosion contributed strongly to collapses. Usually concrete damage would have to be well advanced before a structure is at risk in conventionally reinforced concrete structures.

Particular problems arise when the corrosion problem is the black rust described in Chapter 2 and prestressed, post-tensioned structures where corrosion is difficult to detect as the tendons are enclosed in ducts. Tendon failure can be catastrophic as they are loaded to 50% or more of their ultimate tensile strength and modest section loss leads to failure under load. The particular problems of assessing prestressed concrete structures are addressed in Chapter 4. The Morandi Bridge collapse in 2018 in Italy appears to be a case of a catastrophic failure of corroded post tensioned steel cable stays.

The most common problem caused by corrosion is spalling of concrete cover. A man was killed in New York City by a slab of concrete which spalled off a bridge substructure due to deicing salts and a small truck was badly damaged in Michigan in another similar incident. Special metal canopies have been built around the lower floors of high rise buildings where corrosion has led to risk of falling concrete. This action also allowed the

Figure 3.6 The M4 elevated section in West London showing exposed reinforcement due to concrete spalling induced by reinforcement corrosion. Corrosion was caused by deicing salt leakage from the deck above.

investigators to collect the fallen concrete at regular intervals and weigh it. In that way they could determine whether the corrosion rate was stable, accelerating or decelerating.

A typical example of a corroding substructure is shown in Figure 3.6 where a regular hammer survey was undertaken to remove loose concrete to avoid the risk of it falling on vehicles or pedestrians below until repairs were completed.

The important factor in corrosion of steel in concrete compared to most other corrosion problems is the volume of oxide and where it is formed. A dense oxide formed at high temperatures (such as in a power station boiler) usually has twice the volume of the steel consumed. In most aqueous environments the excess volume of oxide is transported away and deposits on open surfaces within the structure. In the case of steel in concrete two factors predominate. The main problem is that the pore water is static and there is no transport mechanism to move the oxide away from the steel surface. This means that all the oxide is deposited at the metal/oxide interface. The second problem is that the oxide is not dense. It is very porous and takes up a very large volume, up to ten times that of the steel consumed when the porosity of the corrosion product is taken into account. The relative volumes of fully dense oxides are given in Figure 2.2.

The thermodynamics of corrosion, coupled with the low tensile strength of concrete, mean that the formation of oxide breaks up the concrete. It has been suggested that less than 100 micrometres of steel section loss is needed to start cracking and spalling the concrete. The actual amount needed will depend upon the geometry in terms of cover, proximity to corners, rebar spacing, bar diameter and the rate of corrosion. This is discussed further in

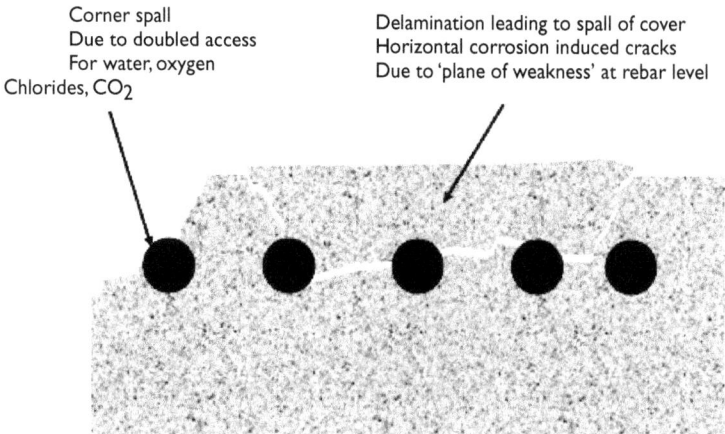

Corner spall
Due to doubled access
For water, oxygen
Chlorides, CO_2

Delamination leading to spall of cover
Horizontal corrosion induced cracks
Due to 'plane of weakness' at rebar level

Figure 3.7 Schematic of corrosion induced spalling at corners and delamination at the plane of the reinforcement.

Section 4.12 on corrosion rate measurement where the amount of corrosion needed to crack concrete is described and in Chapter 9 where the mathematical modelling of the corrosion process is discussed.

Corners tend to crack first on corroding reinforced concrete structures. This is because the oxygen, water, chlorides and carbon dioxide have two faces as pathways to the steel. Delaminations occur as corrosion proceeds on neighbouring rebars and the horizontal cracks join up as shown in Figure 3.7.

3.4 Cracks, crack orientation and corrosion

The importance of cracks in accelerating corrosion by allowing access of corrosion agents to the steel surface has been widely discussed in the literature. If reinforcing steel is doing its job in areas of tension in the structure, small cracks will occur as the tensile load exceeds the tensile strength of the steel. Most of these are small cracks (<0.5 mm) at right angles to the reinforcing steel. They should not significantly affect the rate of corrosion of the steel as any local ingress of chlorides, moisture and carbonation is limited and contained by the local alkalinity. Obviously there is a limit to this 'self healing' ability. If there are large cracks that stay open (>0.5 mm), or excessive shrinkage cracks along the bars then corrosion can be accelerated. The relationship between cracks and reinforcement corrosion is more fully discussed in Concrete Society Technical Report 45, 2015.

Corrosion causes horizontal cracking along the plane of the rebar and the corner cracking around the end rebar. This leads to the loss of concrete cover as shown in Figure 3.7. This is the main consequence of reinforcement corrosion with its subsequent risk of falling concrete and unacceptable appearance.

3.5 The synergistic relationship between chloride and carbonation attack, chloride binding and release

We have already discussed the fact that chlorides can be bound by the concrete. One of the constituents of cement paste is C3A, a complex inorganic aluminium salt. This reacts with chloride to form chloroaluminates. This removes the chloride from availability in the pore water to cause corrosion. This binding process is strongest for chlorides cast into concrete, and is why it was considered acceptable to use sea water to make concrete for many years.

The extent of binding and its effectiveness is not well understood. However, it is known that a reduction in pH as caused by carbonation will break down the chloroaluminates. This leads to a 'wave' of chlorides moving in front of the carbonation front. Consequently structures with chlorides in them that carbonate are more susceptible to corrosion than

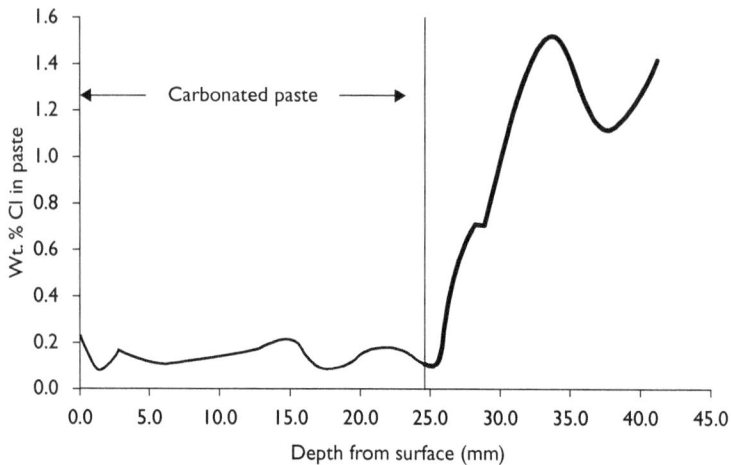

Figure 3.8 A chloride profile through a partially carbonated concrete sample.

those with only one source of problem. A clear example of the unbinding process is shown in Figure 3.8.

References

Concrete Society (2015). The Relevance of Cracking in Concrete to Corrosion of Reinforcement 2nd edition. Concrete Society Technical Report 54. The Concrete Society, Camberley, UK.

Hausmann, D.A. (1967). 'Steel Corrosion in Concrete: How Does it Occur?' *Materials Protection*, 6: 19–23.

Kropp, J. and Hilsdorf, H.J. (1995). Performance Criteria for Concrete Durability Rilem Report 12. Product of RILEM Technical Committee 116-PCD. E&FN Spon, London.

Simon, P. (2004). Improved Current Distribution Due to a Unique Anode Mesh Placement in a Steel Reinforced Concrete Parking Garage Slab CP System. NACE Corrosion 2004. Paper No. 04345.

Trend 2000. BRE; J Broomfield Consultants, and Risk Review Ltd. Evaluation of Life Performance and Modelling. Corrosion of Steel in Concrete – Report D29. 2001 Dec. DTI DMI 5.1 (Report 2).

Vassie, P.R. (1987). The Chloride Concentration and Resistivity of Eight Reinforced Concrete Bridge Decks after 50 years Service. Transport and Road Research Laboratory. Research Report 93.

Woodward, R.J. and Williams, F.W. (1988). 'Collapse of Ynys-y-Gwas Bridge, West Glamorgan'. *Proc. Instn. Civ. Engrs*, 1, 84: 635–669.

Chapter 4

Condition evaluation

We have considered the main mechanisms of corrosion in Chapter 2. We have seen that the chemical process is the same regardless of whether the cause is carbonation or chloride attack as described in Chapter 3. But if we are to perform an effective repair we must fully understand the cause and extent of damage or we risk wasting resources with an inadequate or unnecessarily expensive repair. This chapter explains how to evaluate the condition of corroding reinforced concrete structures.

A full evaluation is normally a two-stage process. The preliminary survey should characterize the nature of the problem and give guidance in planning a detailed survey. The detailed survey will confirm the cause and quantify the extent of the problem. The UK Concrete Society Technical Report 26 (Concrete Society, 1984), the Concrete Bridge Development Group Technical Guide 2 (Concrete Bridge Development Group, 2002), the American Concrete Institute Committee 222 reports (American Concrete Institute, 2019) give excellent reviews of how to conduct the surveys. It is also important to undertake a desk study of the available design, construction, and relevant maintenance drawings and reports.

Corrosion is not the only deterioration mechanism in reinforced concrete. Alkali-silica reactivity (ASR), sulphate attack, thurmasite attack, delayed ettringite formation, freeze thaw, thermal movement, settlement and other movement can all lead to concrete damage and their assessment must be included in the surveys.

Some structures may be prone to unusual chemical attack of the steel or the concrete. For example:

- storage vessels can contain liquids that will attack aggregates, cement paste or the steel;
- carbonates in water can attack concrete pipelines and underground structures.

However, we will concentrate on corrosion of atmospherically exposed reinforced concrete structures and elements.

DOI: 10.1201/9781003223016-4

A condition evaluation as described here is not a structural survey. A structural engineer must be consulted if there are concerns about the capacity of the structure either due to corrosion damage or for any other reasons. Any excessive deflection of structural elements, misalignment, impact damage, excessive cracking, loss of concrete or loss of steel section will require a structural evaluation before repairing corrosion damage. This may lead to limiting the amount of break out permitted during repair, adequate time for concrete strength gain after repair and a requirement for propping during the repair process itself. If repairs are to be structural then a structural engineer must design them. It may be necessary to support the live and dead loads on the structure during structural repair, so that the loads will go through the repairs after the support is removed. In some cases structural support is needed to remove significant amounts of concrete. A structural support system for concrete repair is shown in Figure 4.1.

Figure 4.1 Structural support system on the Midland Links Motorway elevated section (Junction 9 M6) prior to repair. The support system is jacked up to take some of the load while concrete repair is undertaken. Conductive coating cathodic protection systems can be seen on subsequent cross heads. Courtesy of Kevin Davies, Corrosion Consultant, taken December 2005.

4.1 Desk study

Where possible the first stage of an investigation should be to study whatever records are available. Ideally this will consist of 'as built' drawings, inspection records, previous test results and previous repair records. These may be detailed and extensive, for instance in the case of major highway bridges. Alternatively they may be non-existent or illegible in the case of older buildings and structures with multiple previous owners. The desk study may reveal areas at risk, and help plan the investigation. Access requirements can be evaluated from the desk study.

4.2 Preliminary survey

This normally involves a visual inspection, probing of cracks and spalls to see their extent, reinforcement cover measurement, possibly a few carbonation measurements, reference electrode measurements and taking samples (sometimes taking broken pieces of concrete rather than coring) for laboratory testing. Particular attention must be paid to safety and structural integrity from concrete spalling or steel section loss. Loose concrete that could cause a risk to the public or users of the structure should of course be removed. If there are inaccessible areas of potentially loose material then a 'make safe survey' may be an urgent next step, possibly using abseiling operatives if no other access can be provided. Other causes of concrete cracking (e.g. ASR, freeze thaw, thermal movement, structural movement, impact damage, etc.) must not be overlooked at this stage.

4.3 The detailed survey

The purpose of the detailed survey is to ensure a cost-effective repair in line with the client's requirements. This is done by accurately defining and measuring the cause, extent and severity of deterioration. In Chapter 9, we will discuss how test measurements may be used to model the deterioration rate, time to corrosion and life cycle costing. We will need to know how much damage has been done and what has caused the damage. Quantities for repair tenders will probably be based on the results of this survey, so a full survey of all affected elements may be required. Alternatively a full visual survey may be required, with a hammer (delamination) survey of all accessible locations. A number of representative areas may be selected for a detailed survey of cover depths, carbonation depths, chloride content or profile, half cell potentials and other techniques described in the following sections of this chapter.

It is usual to produce pro formas for noting down all deterioration, as well as readings and samples taken so that everything can be tied together in the analysis. Weather conditions are also recorded as these can affect some readings. An example of a report drawing of a condition evaluation is given in Figure 4.2.

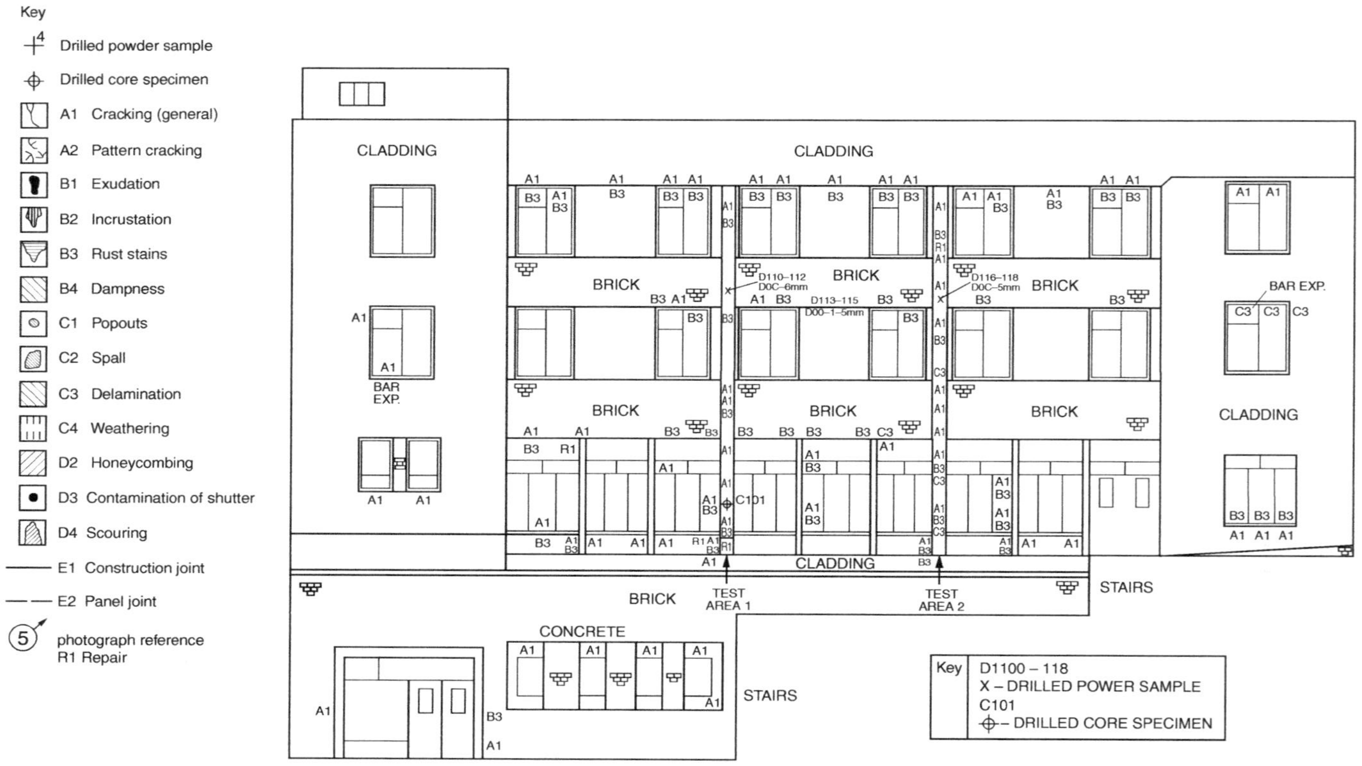

Key

+ Drilled powder sample
⊕ Drilled core specimen
A1 Cracking (general)
A2 Pattern cracking
B1 Exudation
B2 Incrustation
B3 Rust stains
B4 Dampness
C1 Popouts
C2 Spall
C3 Delamination
C4 Weathering
D2 Honeycombing
D3 Contamination of shutter
D4 Scouring
E1 Construction joint
E2 Panel joint
⑤ photograph reference
R1 Repair

Figure 4.2 Typical visual survey of a reinforced concrete framed building suffering from sea salt spray ingress on top of cast in chlorides leading to chloride-induced corrosion.

4.4 Available techniques

The following sections explain some of the available techniques, their advantages and limitations, and the resources needed to employ them. The list is not exhaustive and new techniques are always in development. If all the techniques available were used to thoroughly survey a corroding reinforced concrete structure then huge resources could be used before doing anything to stop the problem. However, expensive repairs can be useless if the problem is not properly diagnosed. So what techniques should be used and what are their limitations and capabilities?

Table 4.1 lists most of the techniques that are used for condition surveying and have relevance to corrosion or the damage it causes. A minimum require-

Table 4.1 Methods for condition surveying

Method	Detects	Use	Approximate speed[a]
Visual	Surface defects	General	$1\,m^2\,s^{-1}$
Hammer/chain	Delaminations	General	$0.1\,m^2\,s^{-1}$
Cover meter	Rebar depth and size	General	1 reading in 2–5 min 50–500 m² per day
Phenolphthalein	Carbonation depth	General	1 reading in 5 min 80–250 per day
Chloride content	Chloride corrosion	General	Core in 10 min 7–80 per day (small or large diameter) or drillings in 2 min 100–250 per day + lab or special site analysis
Reference electrode (half cell) survey	Corrosion risk	General/ specialist	1 reading in 5 s after connecting to steel 50–500 per day
Linear polarization	Corrosion rate	General/ specialist	1 reading in 5 to 30 min depending on equipment used
Resistivity	Concrete resistivity/ corrosion risk	General/specialist	1 reading in 20 s 10 to 20 per day.
Permeability (ISAT)	Diffusion rate	General/specialist	1 reading in 5 min or core + lab 6–12 per day
Impact/ ultrasonics	Defects in concrete	Specialist	1 reading in 2 min 20–30 per day
Petrography	Concrete condition, etc.	General	Core in 10 min 7–80 per day (small or large diameter) + lab
Radar/ radiography	Defects, steel location, condition	Specialist	>1 m² per sec for vehicle system or 1 m² in 20 s for hand system + interpreting

Note
a Daily progress rates from Concrete Bridge Development Group Technical Guide 2 (2002).

ment would usually be a visual survey, a delamination survey, carbonation and chloride measurements, and cover depth measurements.

A petrographic analysis of the concrete is also usually required. These will be done in three or more representative or more severely damaged areas. These measurements should tell the engineer the cause of corrosion, the extent to which chlorides or carbonation have depassivated the steel and the extent of existing concrete damage due to alkali silica reaction or frost and susceptibility to future damage due to aggregates used and the extent of voidage.

Reference electrode surveys are now a standard method of showing how corrosion is spreading ahead of the damage it causes. The other techniques are used for specific requirements as described later.

There are a number of specialist survey companies that carry out condition surveys. Some are independent, and some are attached to large contractors or to consultants. They will get analyses done to the relevant national and international standards with suitable quality management systems for the site work. When developing a specification and contract for such surveys it is important that the required outcome is clearly stated in the documentation and understood by the survey company. It is easy to generate large amounts of test data and fail to come to a conclusion about the cause and extent of damage and the best options for repair. Concrete Bridge Development Group Technical Guide 2 (Concrete Bridge Development Group, 2002) gives detailed information and a model specification for inspection and reinstatement of inspection damage, including November 1991 UK costs for inspection tests and reporting.

4.5 Visual inspection

The visual inspection is the first step in any investigation. It may start out as a casual 'look over' that spots a problem and end up as a rigorous logging of every defect seen on the concrete surface.

4.5.1 Property to be measured

The aim of the visual survey is to give a first indication of what is wrong and how extensive the damage is. If concrete is spalling off then that can be used as a measure of extent of damage. In some cases weighing the amount of spalled concrete with time can be a direct measure of the deterioration rate. In most cases loose concrete should be removed to avoid safety problems.

4.5.2 Equipment and use

The main equipment is obviously the human eye and brain, aided with a notebook, pro forma or hand-held computer and a camera or phone.

Binoculars may be necessary, but close inspection is better if access can be arranged. A systematic visual survey will be planned in advance. Many companies that carry out condition surveys will have standardized systems for indicating the nature and extent of defects. These are used in conjunction with customized pro formas for each element or elevation of the structure. It is normal to record date time and weather conditions when doing the survey, also noting visual observations such as water or salt run down, damp areas, etc. Examples are given in Figures 4.2 and 4.3. Digital cameras have made it easier to produce visually interesting reports and to highlight specific problems but in some cases it can be difficult to pin down exact locations and sizes of defects. Meticulous records of photograph locations, times and features must be kept if they are to be of greatest possible benefit.

4.5.3 Interpretation

Interpretation is usually based on the knowledge and experience of the engineer or technician conducting the survey. The Strategic Highway Research Program (SHRP) has produced an expert system (Kaetzel *et al.*, 1994). This guides the less experienced engineer or technician through the different types of defects seen on concrete highway pavements and structures, including alkali-silica reaction, freeze thaw damage and corrosion. There is a detailed discussion of visual surveys of highway bridges in The Concrete Bridge Development Group Technical Guide 2 (Concrete Bridge Development Group, 2002).

4.5.4 Limitations

The main limitation is the skill of the operative. Some defects can be mistaken for others. When corrosion is suspected it must be understood that rust staining can come from iron bearing aggregates (iron pyrites) as well as from corroding reinforcement. Different types of cracking can be attributed to different causes.

4.5.5 Standards and guidance

The recognition of ASR is discussed in the SHRP manual on ASR (Stark, 1991) and in HWYCON 1994. Visual surveys must always be followed up by testing to confirm the source and cause of deterioration. Concrete Bridge Development Group (2002) has a table of 25 examples of cracks and visual defects with photographic examples in an appendix. There is also a comprehensive guide to all concrete defects in ACI 2008.

4.6 Delamination

As corrosion proceeds the corrosion product formed takes up a larger volume than the steel consumed. This builds up tensile stresses around the rebars.

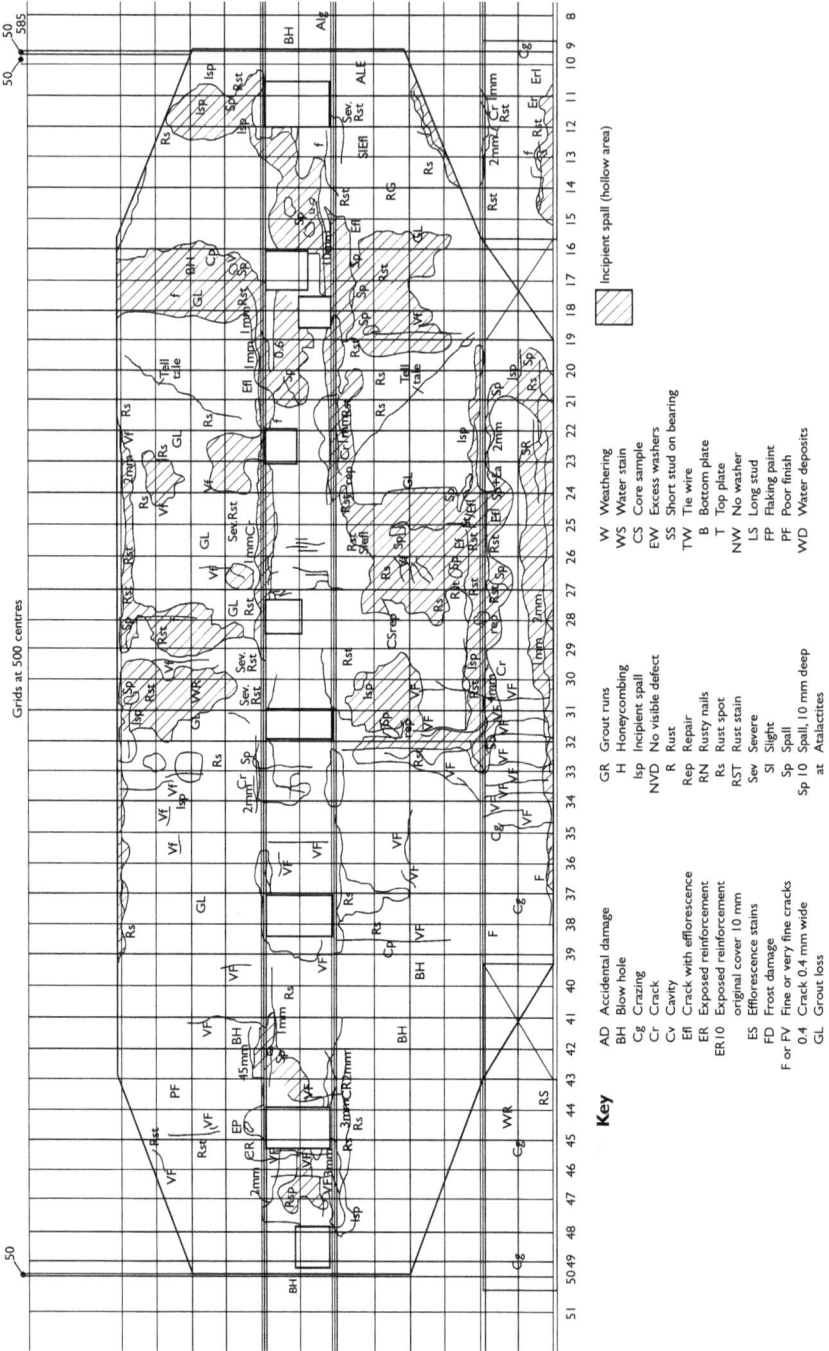

Figure 4.3 Visual and delamination survey of a cross beam on a motorway bridge suffering from deicing salt ingress (developed view of beam). Courtesy of the National Highways.

A layer of corroding rebars will often cause a planar fracture at rebar depth, prior to concrete spalling off as shown in Figure 3.7.

4.6.1   Property to be measured

The aim is to measure the amount of cracking between the rebars before it becomes apparent at the surface. It should be noted that this can be a very dynamic situation. Figure 3.7 shows the way that cracks propagate between corroding bars. The horizontal cracks are detected by a hollow sound when

(a)

(b)

Figure 4.4 (a) Vehicle mounted ground penetrating radar conducting a bridge deck and pavement survey. Courtesy of Penetradar Corp. (b) Infrared thermogram of a patched bridge deck showing a delaminated patch. Reproduced with permission from Transportation Research Board (Strategic Highway Research Program).

hit or other properties associated either with the presence of an unexpected boundary in the concrete or the phase change from concrete to air to concrete when subjected to electromagnetic radiation or ultrasound.

4.6.2 Equipment and use

The hammer survey (or chain drag used on decks) is usually quicker, cheaper and more accurate than the other more sophisticated alternatives such as radar, ultrasonics or infrared thermography. However, these techniques do have their uses, for instance in large scale surveys of bridge decks (radar and infrared) of waterproof membranes or other concrete defects (ultrasonics and radar). A vehicle mounted radar system is shown in Figure 4.4a. The more sophisticated techniques may be needed for deep delaminations and are discussed later in this chapter.

The delamination survey with a hammer is often conducted at the same time as the visual survey as shown in Figure 4.3. Hollow sounding areas are marked directly onto the surface of the structure with a suitable permanent or temporary marker and then recorded on the visual survey pro forma. An automated device was developed by Texas Transportation Institute for highway bridge decks (Gannon and Cady, 1992). The delamatect® used an automated tapper and transducer to detect irregular responses symptomatic of delaminations. These were plotted on a strip chart recorder driven by the turning wheels. It was developed in 1973 and has not been used much after initial trials gave mixed results and the transferring of data from the chart recorder to the deck or a suitable plan of the deck.

Suitably tuned infrared cameras can be used to detect the temperature difference between solid and delaminated concrete. This is best done when the concrete is warming up or cooling down as the delaminated concrete heats and cools faster. This means that the technique's sensitivity is dependent upon the weather conditions and the orientation of the face being surveyed. Infrared thermography tends to work best on bridge decks in the early morning or late evening, or just after dark to eliminate the effect of reflected sunlight. Figure 4.4b shows a bridge deck thermogram. The best systems incorporate a visual light camera for joint recording of visual and infrared image so that the location of defects is recorded simultaneously. The development of global positioning systems makes defect location much easier for vehicle mounted systems. ASTM D4788-03 (2003) gives a standard test method for its application to bridge decks, which can be used as guidance for its application on other concrete elements.

Radar records changes in the dielectric constants associated with the concrete/air phase change. However, the radar also senses the dielectric changes at the steel concrete interface, the presence of water and, to a small amount, chlorides. This makes interpretation of radar images a difficult process. In North America the main use of radar and infrared has been for bridge deck surveys with vehicle mounted systems. In Europe and the

United Kingdom hand-held systems have been used for surveys of building and other structures. The reader is recommended to review the literature for further information (Cady and Gannon, 1992; Bungey, 1993; Titman, 1993; Concrete Society, 1997b; Matthews, 1998; ASTM D6087-05, 2005). There is further discussion of this topic in Section 4.15.

Radar is used in North America for surveying bridge decks using truck mounted rapid data acquisition. They are not accurate in defining the size and location of individual delaminations but can be used for generalized condition surveys or comparative ranking of damaged decks (Alongi et al., 1993). Radar and infrared have been used in combination in North America which increases the accuracy of the measurement of delaminations and other defects but at the expense of doubling equipment and interpretation costs. There is discussion of hand-held radar units in Section 4.15.

4.6.3 Interpretation

The interpretation of radar and infrared is a specialist process usually carried out by the companies who have the equipment and who are hired in to conduct such surveys.

With the hammer or chain drag survey the experience of the operative is vital. A skilled technician who is experienced in carrying out delamination surveys will often produce better and more consistent results than the more qualified but less experienced engineer.

4.6.4 Limitations

The trapping of water within cracks, deep cracks (where bars are deep within the structure) and heavy traffic noise can complicate the accurate measurement of delamination for hammer techniques, radar and infrared thermography.

It is common during concrete repairs for the amount of delamination to be far more extensive than delamination surveys indicate. This is partly due to the inaccuracy of the techniques available but also because of the time between survey and repair. Once corrosion has started delaminations can initiate and grow rapidly. An underestimate of 40% or more is not unusual and should be borne in mind when budgeting for repairs.

Radar has been found to be reasonably accurate in predicting the amount of damage on a bridge deck, but not the precise location of the damage. The problem with thermography is getting the right weather conditions to carry out a useful survey. Radar and thermography have been used in tandem in some North American highway departments with greater success than with individual techniques.

4.6.5 Standards and guidance

ASTM D4580-03 (2003) describes delamination detection in concrete bridge decks using a chain drag or hammers, the delamatect system and a rotary percussion device. The standard concludes that the chain drag procedure is the most reliable technique.

There is a standard, ASTM D4788 (2003), covering infrared thermography of concrete and asphalt-covered concrete bridge decks. This requires a scanner with a minimum thermal resolution of 0.2°C and says that the temperature difference between a sound area and a delamination or a debonded area should be 0.5°C. The standard is for a vehicle mounted system and claims 80–90% of delaminations can be found in concrete decks with or without an asphalt overlay.

ASTM D6087-05 (2005) covers the use of ground penetrating radar to evaluate asphalt covered concrete bridge decks. It is designed for vehicle mounted or manually driven systems. It claims that the system is accurate in detecting delaminations to within ±11.2% according to a precision test on 10 bridge decks in New York, Virginia and Vermont.

Cady and Gannon (1992), Bungey (1993), Titman (1993) Matthews (1998) and Concrete Society (1997b) provide information on radar and infrared.

4.7 Cover

Cover measurement is carried out on new structures to see that adequate cover has been provided to the structure in line with design requirements. It is also carried out when corrosion is observed because low cover will increase the corrosion rate both by allowing the agents of corrosion (chlorides and carbonation) more rapid access to steel, and also allowing more rapid access of the 'fuels' for corrosion, moisture and oxygen. A cover survey will help to explain why the structure is corroding and show which areas are most susceptible to corrosion due to low cover.

4.7.1 Property to be measured

A cover survey requires the location of the rebars three dimensionally, that is, their position with regard to each other and the plane of the surface (X, Y) and depth from the surface (Z). If construction drawings are not available then it may be necessary to measure the rebar diameter as well as its location.

4.7.2 Equipment and use

Magnetic cover meters are now available with logging features and digital outputs. A spacer can be used to estimate rebar diameter. Other alternatives such as radiography can be used to survey bridges or other structures but this is rarely cost effective (Cady and Gannon, 1992; Bungey, 1993). Radar can be employed where the reinforcement layout is complex and congested. Cover meters are surprisingly difficult to use as they are slow and deep cover and closely spaced bars affect the readings. A typical device is shown in Figure 4.4a. An alternating magnetic field is used to detect the presence of magnetic materials such as steel rebars. Modern units are available that will detect non-magnetic stainless steels as well as conventional magnetic steels.

A more recent development is the scanning cover meter which can produce a plot of steel layout and depth. Figure 4.5 shows the different equipment and outputs.

Figure 4.5 (a) A small head manual cover meter being used to detect a corroding wrought iron band in a Grade II listed building. (b) A scanning cover meter being used to map reinforcement layout. Courtesy Hilti Great Britain. (c) Output of a scanning cover meter showing a bar termination. Courtesy Hilti Great Britain.

4.7.3 Interpretation

One of the few standards for cover meters is BS 1881 Part 204. This refers to the measurement on a single rebar. A useful paper has been published which discusses cover meter accuracy when several rebars are close together (Alldred, 1993). The paper suggests that different types of head are more accurate in different conditions. The smaller heads are better for resolving congested rebars.

4.7.4 Limitations

The main problem with cover measurements is the congestion of rebars giving misleading information (Alldred, 1993). Iron bearing aggregates can lead to misleading readings as they will influence the magnetic field. Different steels also have different magnetic properties (at the extreme end, austenitic stainless steels are non-magnetic). Most cover meters have calibrations for different reinforcing steel types. The devices are slow and are not very accurate in the field as any operative who has tried to use one to locate steel and excavate the steel will tell you. You often miss the steel that is indicated when trying to find it to make electrical connections or visually examine it. It is always advisable to check cover meter measurements by breaking out and examining one or more reinforcing bars.

   Problems can arise when the steel is not vertical or horizontal as expected. Scanning cover meters can improve our understanding of what steel lies below the surface. This can be particularly important when designing cathodic protection systems as discussed in Section 7.3.11.

4.7.5 Standards and guidance

One of the few standards for cover meters is BS 1881 Part 204. Alldred (1993) discusses cover meter accuracy when several rebars are close together. Concrete Bridge Development Group Technical Guide 2 (2002) gives good coverage of cover meters, their design and performance.

4.8   Reference electrode (half cell) potential measurements

The electrochemistry of corrosion, cells and half cells were discussed in Section 2.4. The standard reference electrode or half cell is a simple device. It is a piece of metal in a fixed concentration solution of its own ions (such as copper in saturated copper sulphate, silver in silver chloride, etc.). If we connect it to another metal in a solution of its own ions (such as iron in

$Fe(OH)_2$, see equations 1.1 and 1.3) there will be a potential difference between the two 'half cells'. We have made a simple electrical cell like the Daniell cell in Figure 2.5. The cell will generate a voltage because of the different positions of the two metals in the electrochemical series (Table 2.1) and due to the difference in the solutions (Figure 2.6). This is galvanic action in which a voltage is developed, corrosion occurs and current flows due to coupling of different metals. A second type of cell is a concentration cell that will generate a voltage depending upon differences in the concentration of the solution (strictly the activity), around otherwise similar electrodes.

4.8.1   Property to be measured

By keeping one half of the cell standard (our reference electrode or half cell) and moving it along the concrete surface, we change our full cell by the difference in condition of the steel surface below the moving reference electrode. If the steel is passive the potential measured is small (typically 0 to −200 mV, or even a small positive reading against a copper/copper sulphate electrode); if the passive layer is failing and increasing amounts of steel are dissolving (or if small areas are corroding but the potential is being averaged out with passive area), the potential moves more negative, typically −350 mV. At more negative than −350 mV the steel is usually corroding actively. By convention, we connect the positive terminal of the voltmeter to the steel and the negative terminal to the reference electrode. This gives a negative reading for a silver/silver chloride or copper/saturated copper sulphate reference electrode.

Very negative potentials can be found in saturated conditions where there is no oxygen to form a passive layer but with no oxygen there can be no corrosion (see Section 2.2). This shows the weakness of potential measurements. It is a measure of the thermodynamics of the corrosion, not of the rate of corrosion. Corrosion potentials can be misleading, their interpretation is based on empirical observation, not rigorously accurate scientific theory. The problem is that the potential is not purely a function of the corrosion condition but also other factors, and that the corrosion condition is not the corrosion rate.

The reference electrode potential measurement gives an indication of the corrosion risk of the steel. The measurement is linked by empirical comparisons to the probability of corrosion.

The major factors that affect the activity of the iron in solution are the extent to which the steel is depassivated, that is, the extent of carbonation around the steel or the presence of sufficient chloride to break down the passive layer, and the presence of oxygen to sustain the passive layer. In the absence of oxygen, iron will dissolve but will remain stable in solution as there is no compensating cathodic reaction so the potential may be very negative but the corrosion rate will be low. However, there may be a very large (negative) potential against a standard reference electrode.

4.8.2   Equipment and use

The equipment used is illustrated schematically in Figure 2.6. It consists of a voltmeter, reference electrode and cables. A high impedance digital voltmeter is required (around 10 megohm). This is connected to the reinforcing steel and to the reference electrode. Silver/silver chloride/potassium chloride (Ag/AgCl/KCl) and manganese/manganese dioxide/sodium hydroxide (Mn/MnO$_2$/NaOH) double junction electrodes are recommended. These use gel electrolytes and are of very low maintenance. Copper/copper sulphate (CSE) cells are also used and now come in a gel electrolyte form. CSE electrodes with a liquid saturated copper sulphate electrolyte are not recommended because of the maintenance needs, temperature and photosensitivity, the risk of contamination of the cell, the difficulty of use in all orientations and the potential for leakage of corrosive copper sulphate.

It is important to record the equipment used. Different reference electrodes have different 'offsets'. The Ag/AgCl/KCl electrode gives potentials that are a function of the potassium chloride concentration in them. This is usually about 110 mV more positive than a copper/saturated copper sulphate electrode. This can be compensated for internally in the logging equipment if used or during reporting if the ASTM criteria are being used (discussed later).

Standard electrode potentials are given against the 'hydrogen scale'. That is against a standard cell consisting of one atmosphere (strictly unit fugacity) of hydrogen gas in a solution containing one mole (strictly unit activity) of hydrogen ions. The cell itself has a platinum electrode with hydrogen bubbling over it and is in a 1M solution of hydrochloric acid.

It is more usual to measure potentials in the laboratory against a saturated calomel electrode (mercury in saturated mercuric chloride). This cell is recommended for calibrating field reference electrodes. It is possible to adjust the chloride content of a silver/silver chloride cell so that it behaves like a calomel cell. A calomel cell is not usually used in the field because it contains mercury.

A high impedance digital volt meter is used to collect the data in the simplest configuration. Other options are to use a logging voltmeter (or logger attached to a voltmeter), an array of cells with automatic logging or a reference electrode linked to a wheel for rapid data collection (Broomfield *et al.*, 1990). An example is shown in Figure 4.6.

Loggers can be linked to individual reference electrodes or built into the wheel or array systems as in Figure 4.6. They will store readings against

Figure 4.6 A 'potential wheel' scanning a steel column embedded in a brick facade to detect corrosion risk.

position data. They will download to computers or tablets to view the data on site.

The procedure of measurement with a reference electrode is as follows:

1 Check and calibrate the reference electrode.
2 Check the digital voltmeter or logger and the associated cables and connections.
3 Select the area of measurement, usually a whole element such as a bridge deck or cross head beam, or representative areas of element or structures several square metres in size.
4 Use a cover meter to locate the steel and determine rebar spacing.
5 Make a connection to the steel either by exposing it or using already exposed steel. The connection must be metal to metal and secure. Self-tapping screws are frequently used.

6   Check that the steel is continuous with a digital volt meter between two points that are well separated and on well separated rebars. The resistance should be less than 1 ohm or the DC voltage reading should be less than 1 mV. Readings should be stable and should not change significantly when the connections are reversed.

7   Mark out a grid. This will typically be 0.2–0.5 m² but may be smaller, larger or rectangular depending on the steel spacing, the geometry of the element being surveyed and other factors determined by the experience of the investigator. The grid may coincide with the rebar spacing on small surveys, but not usually on larger scans.

8   If necessary wet the area to ensure good contact between the electrode and the concrete. Alternatively, wet the immediate area of the measurement. Tap water, soap solution and even saline solutions have been recommended for wetting. The author prefers tap water. It may be necessary to chip away surface contaminants or coatings. The surface should be damp, not flooded. For a reading to be made, ionic contact is required between the reinforcing steel and the metal in the reference electrode. The concrete must be damp enough for ions to flow. Direct contact between the reference electrode and the steel must not occur.

9   Record the environmental conditions, details of the reference electrode, contact fluid, electrical connection, reinforcement cover depth, condition of the concrete and the precise location of the measurements. These factors may affect readings and their interpretation. If there is a concern about local DC power supplies the issue of stray current may require addressing.

10 Take and record the readings. For manual readings (without logging equipment), it is good practice to take two immediately adjacent readings to check that they are within a few millivolts of each other.

11 Examine for anomalies, check most negative reading areas for signs or causes of corrosion, usually by systematically breaking out, examining and recording the condition of reinforcement.

Data are normally recorded on a plan reflecting the survey grid. The interpretation and presentation is discussed later. The potential map should be drawn up while still on site in order to check that the data are sensible and that apparent 'corrosion hot spots' are investigated as part of the survey.

4.8.3   Interpretation and the ASTM criteria

The main standard for measuring and interpreting reference electrode potentials on reinforced concrete is ASTM C876. However, there are also guidance documents published by Rilem (Elsener, 2003), The Institute of Civil Engineers (Chess and Gronvold, 1996) and the Concrete Society (2004). C876 has an appendix, X1, which defines high, indeterminate and low corrosion risk potentials vs CSE. ASTM quotes steel potentials values against a copper/copper sulphate reference electrode. However, as noted earlier, copper/copper sulphate is not recommended and cells should be calibrated against a calomel cell. Therefore, the criteria are given in Table 4.2 against a saturated calomel electrode (SCE), the standard hydrogen electrode (SHE) and a routinely used silver/silver chloride electrode. It should be noted that reference electrode potentials have been found to be stable and reproducible ±25 mV. Also there are some discrepancies in the literature as to the exact correspondence between different electrode types.

The negative sign is by convention and will depend upon how the leads are connected to the reference electrode and the rebar from the millivoltmeter.

This interpretation was devised empirically from salt-induced corrosion of cast-in-place bridge decks in the United States. Also offsets are sometimes seen for different types of concrete (precast or containing cement replacement materials).

Problems with this interpretation occur for a number of reasons. When there is little oxygen present, especially where saturated with water, the potentials can go very negative without corrosion occurring. The wet bases of columns or walls often show more negative potentials regardless of corrosion activity. Very negative potentials have been measured below the water line in marine environments, however, the lack of oxygen will often slow the corrosion rate to negligible levels. Other problems may arise in the presence of carbonation, high-resistivity concrete and with stray electrical currents.

In carbonated concrete the anodes and cathodes are so close together that a 'mixed potential' (an average of the anode and cathode) is measured. Also carbonated concrete wets and dries quickly as the pores are partly blocked by the calcium carbonate deposits. This means that the resistivity of the concrete will affect the measurement. If the reading is taken with no wetting a very positive reading may be found. If the reading is taken after wetting the measurement may drift more negative for many hours.

Table 4.2 ASTM criteria for corrosion of steel in concrete for different standard reference electrodes

Copper/copper sulphate	Silver/silver chloride/ 1.0M KCl	Standard hydrogen electrode	Calomel	Corrosion condition
>−200 mV	>−100 mV	+120 mV	>−80 mV	Low (10% risk of corrosion)
−200 to −350 mV	−100 to −250 mV	+120 to −30 mV	−80 mV to −230 mV	Intermediate corrosion risk
<−350 mV	<−250 mV	−30 mV	<−230 mV	High (>90% risk of corrosion)
<−500 mV	<−400 mV	−180 mV	<−380 mV	Severe corrosion

A third problem arises due to the existence of the carbonation front. This is a severe change in the chemical environment from pH 12 to pH 8, that is, a factor of 10^4 difference in the concentrations of the hydroxyl and hydrogen ions, with similar changes in the calcium and other metal ions that precipitate out on carbonation. This can lead to a 'junction potential' superimposed on the corrosion potential giving misleading results (Arup and Klinghoffer, 1998).

For carbonated concrete one method is first to do a potential survey with minimum wetting (if stable potentials can be established). Then wet the surface thoroughly and leave it for at least two hours or until potentials are stable. Resurvey once potentials have stabilized and then look for the most anodic (negative potential) areas. These are most probably active if a potential difference of 100 mV or more exists over a space of 1.0 m or less. A physical investigation is essential to see if there is reasonable correlation between corrosion and anodic areas.

Given the thermodynamic nature of the measurement and our current understanding of potential measurements, that is probably the best we can do presently on interpretation of reference electrode potentials. Corrosion rate measurement is discussed in Section 4.12.

Stray currents can also influence the readings. These were discussed in Section 2.3. The effect of stray currents on reference electrode potential readings can be used as a diagnostic tool where stray current corrosion is suspected in the presence of DC fields. If a reference electrode is mounted in or on the concrete and attached to the reinforcement via a logging voltmeter, fluctuations in the potential may be linked to the operation of nearby DC equipment, especially if the equipment can be deliberately turned on and off and the potentials fluctuate accordingly. According to BS7361 Part 1 (1991) the maximum positive change due to stray currents should not exceed 20 mV (there is some discussion in the standard about steel in concrete and the passivation effects but in the end the 20 mV criterion is recommended here too). There is discussion of stray current corrosion and its measurement in AMPP/NACE Report 011101 (2010) and AMPP/NACE SP 21427 (2019).

There has been a tendency to correlate reference electrode potentials with corrosion rate. The reference electrode potential is a mixed potential representing anodic and cathodic areas on the rebar, it is not the driving potential in the corrosion cell. Any correlation between potential and corrosion rate is fortuitous and is often due to holding other variables constant in laboratory tests.

There has been some discussion earlier and in the literature of the 'junction potentials' created by the change in chemical concentrations within the concrete (Bennett and Mitchell, 1992). This effect was severe in a concrete slab subjected to chloride removal, but that may be due to the treatment (discussed in a later chapter), rather than being a real problem under normal conditions. However, the junction potential may explain the erratic changes in potentials seen in carbonated structures as potentials exist across the carbonation front due to the pH change, and due to concentration changes as carbonated concrete wets and dries quickly because of the lining of the pores by calcium carbonate. Reports of potential measurements on structures treated with electrochemical chloride extraction are varied. Some see normal potential ranges quite quickly, others see highly polarized steel (very negative potentials) for periods exceeding a year.

A further cause of error can be the proximity of other metals, particularly galvanizing, which move the potential to more negative values (often see on buildings around galvanized steel window frames).

A histogram or cumulative frequency plot will show what proportion of measurements exceed the criteria to show the extent of high corrosion risk. Where the ASTM criteria do not apply they will show the distribution of readings so that high risk areas can be identified.

The best way of interpreting reference electrode potential data is to expose areas of rebar which show the most negative potentials, intermediate and least negative potentials to correlate corrosion condition with readings. If there are severe potential gradients across the surface then corrosion is likely to be localized with pitting present. Care must be taken in areas of moisture, for instance, puddles that remain on the surface or where moisture comes up from the ground. These may show very negative potentials due to chloride accumulation or to oxygen starvation. Corrosion rate measurements may be required to determine whether a very negative potential measurement is an artifact or to a high active level of corrosion.

4.8.4 Reference electrode potential mapping

A fuller understanding of the corrosion condition is given by drawing a potential map of the area surveyed. This is a plot of the readings where lines are drawn separating the levels of potential. This shows the high corrosion risk areas and the low corrosion risk areas. A rapid change in potential is seen as a steeper gradient. This indicates a greater risk of corrosion. Figure 4.7 shows a typical reference electrode potential plot for a bridge beam, which correlates with the visual survey in Figure 4.3. Figure 4.8 shows a survey of a leaf pier with chlorides leaking from the deck above and being splashed up from the road traffic. While these isopotential contour plots are not as quantitative as simply following the ASTM criteria, such mapping of anodic areas is valid over a wider range of structures and conditions.

A line of potential measurements can be plotted on a distance vs. potential plot. This will show which points exceed the ASTM criteria and where the steepest potential gradients and the most anodic areas are with the most negative values.

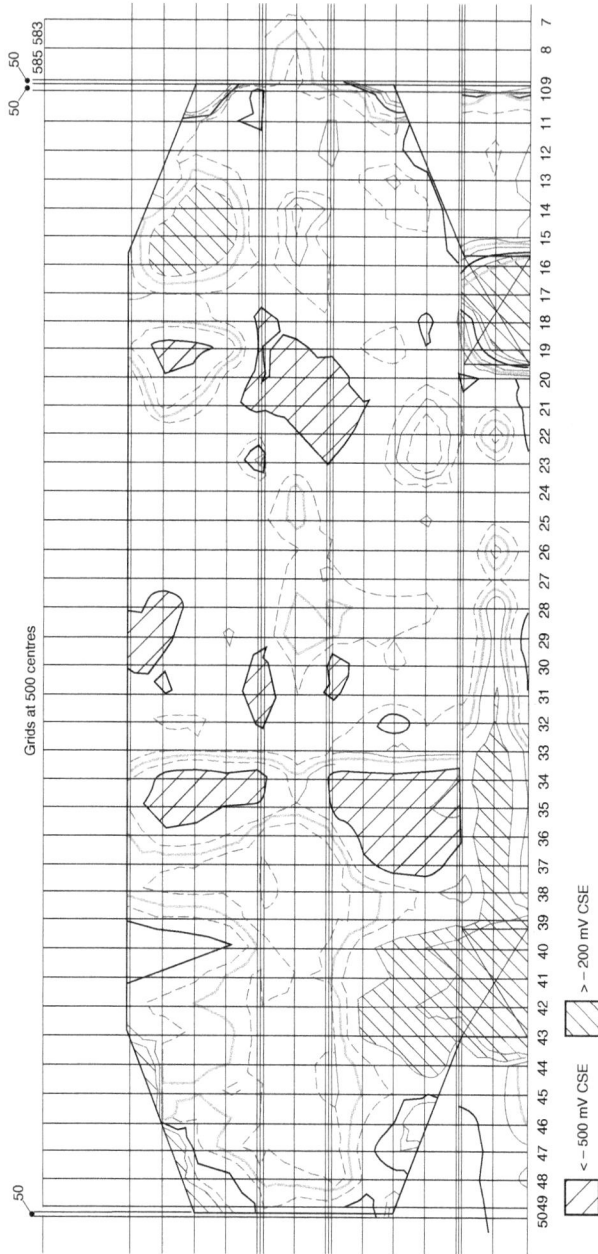

Figure 4.7 Potential survey as per delamination and visual survey. Figure 4.3. Courtesy of National Highways.

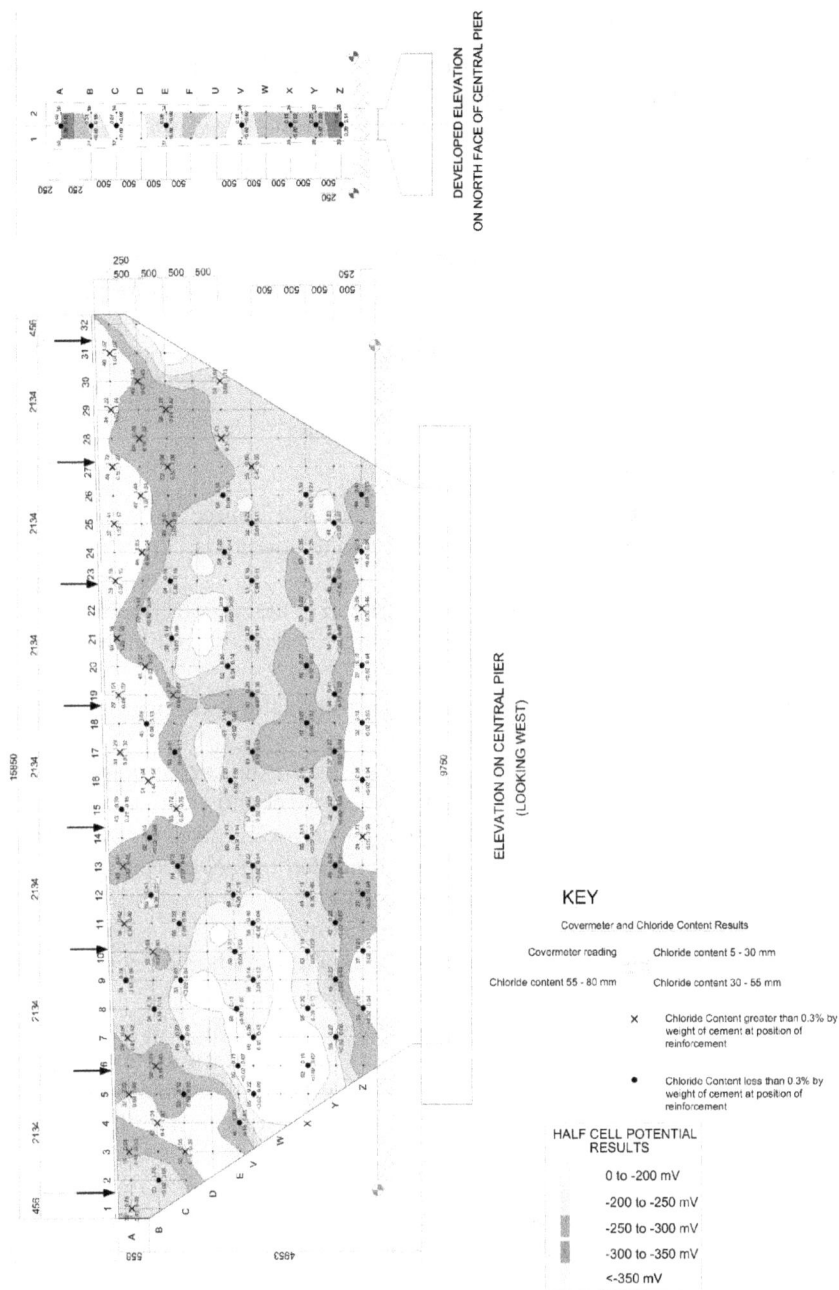

Figure 4.8 Reference Electrode Potential survey of a Bridge Pier, also recording cover meter chloride concentrations at three different depths. Courtesy National Highways.

4.8.5   Cell to cell potentials

If it is not feasible to make direct connections to the reinforcing steel it is possible to get comparative potential data by measuring the potentials between two reference electrodes, with one kept in a fixed position and the other moved across the surface. This is the linking of two full cells, one kept constant (the fixed cell and the steel directly below it), while the other reference electrode moves, changing the steel to concrete reference electrode. We will not know the absolute value to the steel to concrete reference electrode, but we will measure how it changes from point to point. Interpretation of data from fixed vs. moving reference electrode surveys is more difficult than interpreting absolute values, but again, the contour plot is probably the most valid method of interpretation.

4.8.6   Limitations

The following is a non-exhaustive list of possible sources of error in the taking and interpretation of reference electrode potentials:

1   Electrical continuity – poor metal to metal contact, poor reinforcement continuity within survey area or scanning across a discontinuity such as an expansion joint.
2   Surface contact – failure to remove contamination, coatings or to wet the surface adequately.
3   Cracked and spalled concrete – this can distort the current path or give low readings due to poor contact between steel and electrode.
4   Saturated concrete – as discussed earlier, can give very negative potentials while excluding the oxygen needed to fuel corrosion.
5   Other metals – the presence of galvanizing, galvanic anodes or conduits in the concrete can distort potential measurements.
6   Stray currents – nearby sources of DC such as cathodic protection or DC traction systems can lead to potential shifts.
7   Electrochemical treatments – cathodic protection, electrochemical chloride extraction and electrochemical realkalization are designed to shift the potential of the steel. This effect may be permanent in the case of cathodic protection or temporary but quite long term in the case of the other two techniques.
8   Chemical contamination – anything that inhibits or promotes the anodic or cathodic reactions on the steel surface can distort interpretation of potential measurements. This may include surface applied corrosion inhibitors.
9   Carbonation – the issues of carbonation have been discussed in the previous section.

4.8.7 Standards and guidance documents

Detailed methodologies for undertaking reference electrode potential surveys and their interpretation can be found in Concrete Society (2004), Chess and Grønvold (1996) and ASTM C876-15 (2015). A table of relative reference electrode potentials against the standard hydrogen electrode is given in BS EN ISO 12696.

4.9 Carbonation depth measurement

Carbonation depth is easily measured by exposing fresh concrete and spraying it with phenolphthalein indicator solution. The carbonation depth must then be related to the cover (the average and its variation) so that the extent to which carbonation has reached the rebar can be estimated, and the future carbonation rate estimated.

There is some discussion in the previous chapter as to whether the carbonation front is truly as well defined as the indicator shows it to be (see Figure 3.1(b)) but for most practical considerations it is a reliable technique. Some aggregates can cause problems usually making the colour transition difficult to see. Also very poorly consolidated concrete and concrete underground exposed to dissolved carbonates in the water may not show clearly defined carbonation fronts due to the non-uniform progress of the carbonation front.

Splitting cores to expose a fresh surface should be done carefully to prevent dust from carbonated areas contaminating the uncarbonated surface and vice versa.

4.9.1 Equipment and use

Carbonation is easily measured by exposing fresh concrete and spraying on phenolphthalein indicator as shown in Figure 4.9. This can be done either by breaking away a fresh surface (e.g. between the cluster of drill holes used for chloride drilling as described earlier), or by coring and splitting or cutting the core in the laboratory. The phenolphthalein solution will remain clear where concrete is carbonated and turns pink where concrete is still alkaline.

The best indicator solution for maximum contrast of the pink colouration is a solution of phenolphthalein in alcohol and water, usually 1 g indicator in 100 ml of alcohol/water (50 : 50 mix) or more alcohol to water (BRE Digest 405, 1995). If the concrete is very dry then a light misting with water prior to applying the phenolphthalein will also help show the colour. Care must be taken that dust from drilling, coring or cutting does not get on the treated surface. Other indicators such as thymolphtalein, Alizarin yellow and universal indicator have been used, along with pH meters. However, phenolphthalein is the most reliable, convenient and widely used indicator.

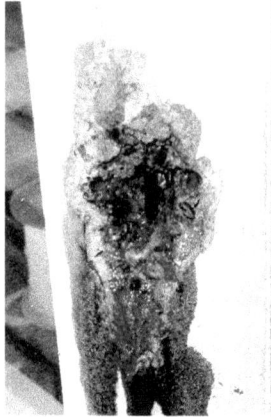

Figure 4.9 Phenolphthalein applied to a concrete window mullion showing approximately 10 mm carbonation depth (clear area at surface) and an uncorroded bar in uncarbonated concrete with a cover depth of approximately 40 mm.

4.9.2 Interpretation

Carbonation depth sampling can allow the maximum, average and standard deviation of the carbonation depth to be calculated. If this is compared with the average reinforcement cover then the amount of depassivated steel can be estimated. If the carbonation rate can be determined from historical data and laboratory testing then the progression of depassivation with time can be calculated. Generally, the maximum carbonation depth at any particular location is the most important. However, if this is at cracks and they are significantly deeper than the general carbonation depth this should be noted.

4.9.3 Limitations

Phenolphthalein changes colour at pH 9. The passive layer breaks down at about pH 11. If the carbonation front is 5–10 mm wide the steel can be depassivated 5 mm away from the colour change of the indicator as shown in Figure 3.1(b). This should be considered when using phenolphthalein measured carbonation depths to determine the rate and extent of depassivation.

Some aggregates can confuse phenolphthalein readings. Some concrete mixes are dark in colour and seeing the colour change can be difficult. Care must be taken that no contamination of the surface occurs from dust and the phenolphthalein sprayed surface must be freshly exposed or it may be carbonated before testing.

It is also possible for the phenolphthalein to bleach at very high pH, for example, after chloride removal or realkalization. If the sample is left for a few hours it will turn pink. There can also be problems on below ground structures where carbonation by ground water does not always produce the clear carbonation front induced by atmospheric CO_2 ingress.

The number of test areas is frequently limited by the amount of concrete that those responsible for the structure will allow to be broken off or holes drilled or cores taken so it is rarely possible to systematically carry out carbonation tests. It is often necessary to carry out carbonation tests away from the public gaze. Often it is not possible to carry out tests systematically. They are often done at accessible locations where damage is not too extensive. Damaged areas should be repaired after coring or breaking out.

4.9.4   Standards and guidance documents

There is a CEN standard on carbonation depth measurement BS EN 14630. The Concrete Society (2004) report is similar to the CEN standard. BRE Digest 405 (1995) specifically discusses carbonation and its measurement in Portland cement concrete. BRE Digest 444 (2000) Part II and BRE IP 11/98 (1998) discuss measurement in high alumina cement concrete (HAC).

4.10   Chloride determination

Chlorides are usually measured by dissolving powder samples in acid. The samples are taken from drillings or from crushed cores. It is preferable to collect a series of drillings at different depths so that a chloride profile can be produced. If a single sample is taken it is recommended that it is taken at the reinforcement depth.

Alternatively a core can be cut into slices and the slices crushed. Like the carbonation tests the chloride profile must be related to cover so that the extent to which rebars are exposed to high chlorides can be determined.

Chloride profiles can also be used to determine the diffusion coefficient and thus predict the ongoing rate of ingress. However, to do this a minimum of four depth increments are required for sensible curve fitting.

The corrosion thresholds were discussed in Section 3.2.3. It is important to recognize that these are approximate for the reasons discussed in Section 3.2.3. It is also important to realize that the chloride level at the rebar determines the present extent of corrosion, but the profile determines the future rate, as that is what drives more chlorides from the environment to the steel surface.

Chloride contents can be measured by several methods. In the laboratory, powdered samples are usually digested in acid and then titrated to find the concentration in the conventional wet chemical method.

In the field there are two well known methods of measuring chlorides: Quantab strips and specific ion electrodes. The former are of modest accuracy and can be made inaccurate by certain aggregate types. The latter can be highly accurate but the equipment is expensive and requires training and a good methodology to use effectively. Any field technique should be checked against laboratory analysis of duplicate samples. In the United Kingdom, laboratory testing to ISO quality management standards is routine and widely available from independent commercial test houses. The cost of laboratory analysis is very competitive. This is not true in many other countries.

The methods discussed above are referred to as the 'total' or 'acid soluble' chloride contents. There are also methods for measuring the 'free chlorides' or water soluble rather than the acid soluble chlorides. This refers to the fact that it is the chlorides dissolved in the pore water that contribute to the corrosion process. Any chlorides chemically bound up in the cement paste (chloroaluminates or C_3A), or bound up in the aggregates are 'background' chlorides that should not contribute to the corrosion threshold.

Unfortunately the water soluble techniques produce results that are difficult to reproduce so they are rarely used in the United Kingdom or Europe, although the 'Soxhlet extraction technique', a method of refluxing concrete chips in boiling water to extract the chloride, is a standard technique used in North America. See, for example, AASHTO T260 'Sampling and Testing for Total Chloride Ion in Concrete and Concrete Raw Materials' which includes a procedure for water soluble ion sample preparation and analysis. This has been adopted by ASTM as ASTM C 1524 Standard test method for water extractable chloride in aggregate (Soxhlet method).

It has also been suggested that the interaction of bound and unbound chlorides is more complex than previously asserted (Glass and Buenfeld, 2000). Bound chlorides may be available and may reduce the time to corrosion initiation. There is one case that the author knows of where a particular aggregate used in Ontario, Canada contains internally bound chlorides that are not available to the cement paste. In this case, an acid soluble test will add those chlorides to those in the cement paste overestimating the corrosion risk. In other cases the use of the acid soluble test is a reasonable method of assessing the corrosion risk. Chlorides can be cast into concrete or can be transported in from the environment. The chloride ion attacks the passive layer even though there is no significant generalized drop in pH. Chlorides act as catalysts to corrosion. They are not consumed in the process but help to break down the passive layer of oxide on the steel and allow the corrosion process to proceed quickly.

Chloride testing will show:

1   Whether chlorides are present in a high enough concentration to cause corrosion (typically above 0.2–0.4% chloride by weight of cement is at risk of corrosion).
2   Whether chlorides were cast in or diffused in later. Distribution of chloride with depth and about the structure will show a profile with

depth if chlorides diffused in. If there is either even distribution, random distribution or one that looks more closely related to the (cold weather) construction schedule than any exposure to external sources of chlorides then chlorides could have been cast in either as a rapid setting agent or contaminated water or aggregates.

4.10.1 Property to be measured

The amount of chloride ion in the concrete can be measured by sampling the concrete and carrying out chemical analysis (titration) on a liquid extracted from the sample. This is usually done by mixing acid with drillings or crushed core samples. An alternative is pore extraction by squeezing samples of concrete or, more usually, mortar. This technique is frequently used in laboratory experimental work as it is often difficult to extract useful pore water samples from field concrete. The 'Soxhlet extraction' techniques for free chlorides was discussed earlier.

Considerable work has gone into differentiating between bound and free chlorides. As only the free chlorides contribute to corrosion that is ideally what we want to know. However, the binding of chlorides is a reversible and dynamic reaction, so attempts to remove and measure free chlorides will release bound chlorides. A further complication is that carbonation breaks down chloroaluminates thus freeing chlorides which proceed as a wave ahead of the carbonation front.

The most accurate and reproducible tests are the acid soluble chloride tests that effectively measure total chlorides. Pore water extraction and water soluble chloride measurement are less reproducible and less accurate but are probably what we really want to know. However, the chloride threshold values are based on total chloride levels so the system is at least self consistent.

4.10.2 Equipment and use

The collection of chloride samples should be done incrementally from the surface either by taking drillings or sections from cores. The first 5–10 mm is usually discarded as being directly influenced by the immediate environment. This first increment can show excessively high levels if salt has just deposited on the surface or excessively low levels if rain or other water has just washed away the chlorides as shown in Figure 3.2(b). Care should be taken to minimize cross contamination of samples at different increments.

Measurements of chloride content are made at suitable increments, typically 5–10 mm but often bigger, up to 25 mm. For improved statistical accuracy when taking drillings, multiple adjacent drillings are made and the depth increments from each drilling are mixed. Special grinding kits are available and sample sizes required for analysis vary from 10 to 50 g (see

Figure 4.10 RCT Profile Grinder used to collect incremental chloride measurements to produce a chloride profile which can be used to calculate the diffusion coefficient for an existing structure as discussed in Section 9.2. Courtesy Germann Instruments.

Figure 4.10). Methods of sampling are discussed in CBDG (2002) and Concrete Society (2004). See also Section 9.2.2.

The major concern with sample size is ensuring that there is a uniform amount of cement paste in each sample and that there is no risk of the sample being dominated by a large piece of aggregate. Some researchers have crushed cored and removed the larger aggregate pieces, measuring only the paste and small aggregates. This is time consuming and the sample is no longer a balanced sample as the removal of aggregates cannot be done quantitatively.

Each sample should be at least 25 g for accurate analysis. Dry drilling two 25 mm diameter holes (typically 30–40 mm apart) to a 15–25 mm depth has been found to give an adequate sample. Care should be taken to collect all the fine powder as this has the highest chloride content (Concrete Society, 2004).

There are several ways of measuring the chlorides once samples are taken. Field measurements of acid soluble chloride can be made using a chloride specific ion electrode (Herald *et al.*, 1992). Conventional titration by BS 1881 part 124 and potentiometric titration methods are also available (Grantham, 1993). The European standard, BS EN 14629, is based on the BS 1881 method and supersedes it. The methodology described in Concrete Society (2004) is based on BS EN 14629. Generally measurements are quoted in increments on 0.01% chloride by mass of sample. This is frequently converted to chloride by mass of cement assuming 14% cement content giving a lowest resolution of 0.07% by mass of cement.

There has been some discussion in the literature about the accuracy and reproducibility of test results (Nustad, 1994; Grantham and Van Es, 1995).

Problems with the accuracy of measurement have led to recommendations to split samples and either submit them separately for analysis as different specimens to check how closely the results come out or submit duplicates to another laboratory for independent checking. Alternatively it is possible to purchase dust samples of known chloride content for inclusion with the unknown samples.

As well as acid soluble chloride there are the water soluble chloride tests (ASTM C 1524; AASHTO T 260). These techniques use different levels of pulverization of large samples that are refluxed to extract the supposedly unbound chlorides. These are the chlorides that are free in the pore water to cause corrosion as opposed to the chloride bound by the aluminates in the concrete, or bound up in some aggregates of marine origin that are also bound up. The water soluble chloride test is rather inaccurate as the bound chlorides can be released and the finer the grinding the more will be extracted. However, this test can be useful in showing the corrosion condition where chlorides have been cast into concrete, and particularly where aggregates are known to contain chlorides that do not leach out into the pore water.

4.10.3 Interpretation

There is a well known 'chloride threshold' for corrosion given in terms of the chloride/hydroxyl ratio as discussed in chapter 3 and in Hausmann (1967). By simple modelling of chloride and hydroxyl ions contacting the steel surface, Hausmann showed that when the chloride concentration exceeds 0.6 of the hydroxyl concentration the passive layer will break down. He corroborated this with laboratory testing in calcium hydroxide solutions. The 0.6 threshold approximates to a concentration of 0.2–0.4% chloride by weight of cement, 1lb/cu yard of concrete or 0.05% chloride by weight of concrete. The threshold is discussed in Section 3.2.3. A figure was produced in BRE Digest 444 (2000) from empirical evidence of chloride thresholds vs. corrosion and shows the probabilistic nature of the corrosion threshold for chlorides in concrete. There is also a discussion on the statistical distributions of chloride thresholds by Lawler et al. (2021).

The American Concrete Institute gives recommendations for chloride limits in the ACI 222 Guide to protection of metals in concrete against corrosion with limits for acid soluble and water soluble chlorides, for prestressing and reinforcing steels and for wet and dry conditions. The UK Highways agency has a limit of 0.3% by mass of cement by the acid soluble BS EN 14629 standard.

Table 4.4 gives the criteria widely used in Europe for existing structures.

This assumes there is sufficient moisture and oxygen for corrosion and that the location observed is not acting as a cathode to a local anode. Figure 3.4(a) and (b) show results from Vassie (1987) from 50-year-old UK

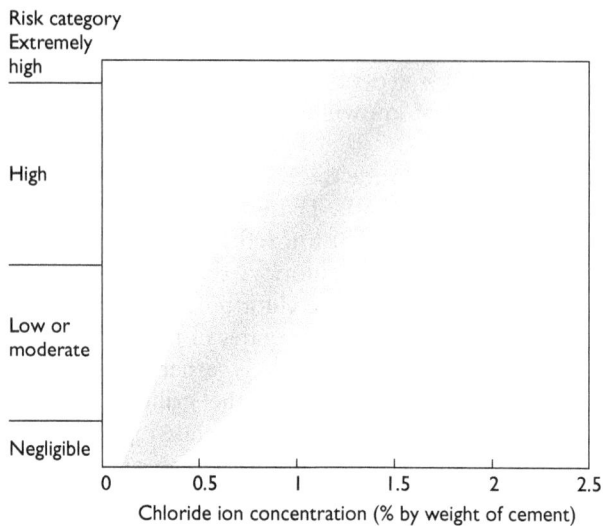

Figure 4.11 Estimated risk of steel reinforcement corrosion associated with ingressed chloride in the absence of carbonation. Reproduced from BRE Digest 444 Part 2. Copyright IHS Markit, reproduced with permission.

Table 4.3 Chloride limits for new construction in % chloride by mass of cement, American Concrete Institute, 2019

Category	Chloride limit for new construction (percent by mass of cementitious material)	
	Test method	
	Acid soluble	Water soluble
	ASTM C1152M	ASTM C1218M
Prestressed concrete	0.08	0.06
Reinforced concrete in wet conditions	0.20	0.15
Reinforced concrete in dry conditions	0.30	0.25

highway bridges. This demonstrates that there must always be cathodes regardless of chloride level so that the probability of corrosion is never 100% even at 2% chloride by mass of cement.

The important questions from chloride measurement are how much of the rebar is depassivated and how will this progress? Points (a) to (c) in Section 3.2.3 review how the corrosivity of the chloride can change. If

Table 4.4 Corrosion risk at given chloride contents

% Chloride by mass of cement	% Chloride by mass of sample (concrete)	Risk
<0.2	<0.03	Negligible
0.2–0.4	0.03–0.06	Low
0.4–1.0	0.06–0.14	Moderate
>1.0	>0.14	High

chlorides have been transported in from outside then the chloride profile can be used along with measurements (or estimates) of the diffusion constant to estimate future penetration rates and the build up of chloride at rebar depth.

Methods of predicting chloride diffusion and the movement of the chloride threshold are discussed in Section 9.2.

4.10.4 Standards and guidance

CEN has produced a test method for chloride analysis. Products and systems for the protection and repair of concrete structures – Test methods – Determination of chloride content in hardened concrete. BS EN 14629.

There are three relevant ASTM standards: ASTM C1152/C1152M Standard test method for acid soluble chlorides in concrete and mortar; ASTM C1524 Standard test method for water-extractable chloride from aggregate; and ASTM C1218/C1218M Standard test method for water soluble chloride in mortar and concrete. ASTM committee G01.14 is also in the process of developing a laboratory test for chloride thresholds for concretes with different admixtures. American Concrete Institute (2019) gives extensive discussion on thresholds as well as the Table 4.4. BRE Digest 444 (2000) has a comprehensive table and 'cloud data' graph of chloride content versus risk based on field data shown in Figure 4.11.

4.11 Resistivity measurement

Since corrosion is an electrochemical phenomenon, the electrical resistivity of the concrete will have a bearing on the corrosion rate of the steel as an ionic current (electric current in the form of a flow of charged ions in the pore water) must pass from the anodes to the cathodes for corrosion to occur.

The four probe resistivity meter or Wenner Probe was developed for measuring soil resistivity (ASTM G57). Specialized modifications of the Wenner probe are frequently used for measurement of concrete resistivity on site. The measurement can be used to indicate the possible corrosion activity if steel is depassivated. A proprietary version of the system is shown in Figure 4.12. Most systems uses four probes. The outer probes pass a current through the concrete while the inner probes detect the voltage difference. This approach eliminates any effects due to surface contact resistances.

For a semi-infinite, homogeneous material the resistivity ρ is given by:

$$\rho = 2\pi a \frac{V}{I}$$

where a is the electrode spacing, I is the applied current across the outer probes and V is the voltage measured between the inner probes.

At one time it was considered necessary to drill holes to insert the probes, but modern four probe devices are spring loaded and just push onto the concrete surface. In one version a wetting gel is applied, in another wooden plugs in the end of the probes are wetted. Cheaper, less accurate two probe systems are also available. These are often inserted into drilled holes in the concrete to improve electrical contact by getting below the surface latence and any minor carbonation.

An alternative approach using a single electrode on the surface and the rebar network can be used to measure the resistivity of the concrete cover. This is available as part of a corrosion rate measuring device (see Section 4.12 and Figure 4.13) and uses the reinforcement cage as one electrode and a small surface probe as the other electrode. The advantage of this approach is that it measures the resistivity of the cover concrete only. The disadvantage is that it suffers from contact resistance problems.

4.11.1 Property to be measured

The electrical resistivity is an indication of the amount of moisture in the pores, and the size and tortuosity of the pore system. Resistivity is strongly affected by concrete quality, that is, cement content, water/cement ratio, curing and additives used. The chloride level does not strongly affect resistivity as there are plenty of ions dissolved in the pore water already and a few more chloride ions here or there does not make a big difference. However, chlorides in concrete can be hygroscopic, that is, they will encourage the concrete to retain water. This is why chlorides are often accused of reducing concrete resistivity.

4.11.2 Equipment and use

Millard (1991) described two versions of the four probe equipment. The suppliers and the details of the equipment have changed over the years but are similar and are shown in more recent reports such as Concrete Society (2004).

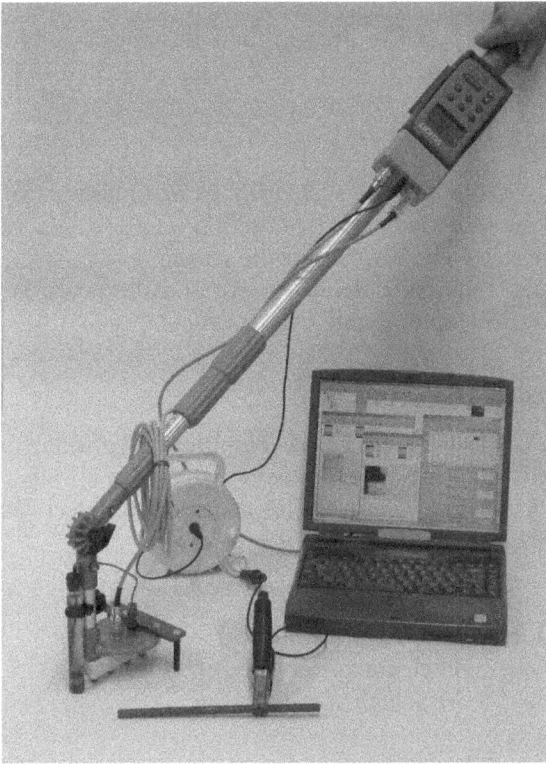

Figure 4.12 A combined four probe resistivity meter with reference electrode and temperature sensor, showing logger, computer, reinforcement connection and sensors on an extension PLO. Courtesy of ClTec GMbH.

It is generally agreed that the four (or two) probe system needs a probe spacing larger than the maximum aggregate size to avoid measuring the resistivity of a piece of aggregate rather than of the paste and aggregate. If it is not possible to avoid the influence of reinforcing steel then measurements should be made at right angles to the steel rather than along the length of it as the steel can provide a 'short circuit' path for the current during the measurement. Measurements should also be taken away from edges of the concrete and water must not be ponded on the surface during readings.

An alternative approach measures the resistivity of the cover concrete by a two electrode method using the reinforcing network as one electrode and a surface probe as the other tubular sensor (Feliú *et al.*, 1988; Broomfield *et al.*, 1993, 1994). The GECOR 6 model that used this technique has now been replaced with GECOR 10 which uses the four probe method.

However, the methodology and the equation can be useful for electrochemical calculations.

Concrete resistivity of the area around the sensor is obtained by the formula:

Resistivity $= 2RD$ $(\Omega \text{ cm})$

where R is the resistance by the 'iR drop' from a pulse between the sensor electrode and the rebar network (see Section 6.5 for a discussion of the iR drop); D is the electrode diameter of the sensor.

This approach requires a damp surface and the probe is best situated between bars rather than directly over them.

4.11.3 Interpretation

Interpretation of resistivity results is empirical. The following interpretation of resistivity measurements from the Wenner four probe system have been cited when referring to depassivated steel (Langford and Broomfield, 1987).

>20 kΩ cm	Low corrosion rate
10–20 kΩ cm	Low to moderate corrosion rate
5–10 kΩ cm	High corrosion rate
<5 kΩ cm	Very high corrosion rate.

Researchers working with a field linear polarization device for corrosion rate measurement have conducted laboratory and field research and found the following correlation between resistivity and corrosion rates using the two electrode surface to rebar approach (Broomfield *et al.*, 1993).

>100 kΩ cm	Cannot distinguish between active and passive steel
50–100 kΩ cm	Low corrosion rate
10–50 kΩ cm	Moderate to high corrosion where steel is active
<10 kΩ cm	Resistivity is not the controlling parameter.

In the previous method and interpretation resistivity measurement is used alongside linear polarization measurements (see Section 4.12), not as a stand alone technique.

4.11.4 Limitations

The resistivity measurement is a useful additional measurement to aid in identifying problem areas or confirming concerns about poor quality concrete. Readings can only be considered alongside other measurements.

There is a frequent temptation to multiply the resistivity by the reference electrode potential and use Ohm's Law to present this as the corrosion current and hence corrosion rate. This is incorrect. The corrosion rate is a function of the interfacial resistance between the steel and the concrete, not the bulk concrete resistivity. The potential measured by a reference electrode on the concrete surface is not the potential at the steel surface that drives the corrosion cell. Correlations between resistivity, reference electrode potential and corrosion rate may be found in similar samples in similar conditions in the laboratory but in the variability of the real world any correlation is fortuitous.

The main problem with the four probe technique is that the reinforcing steel will provide a 'short circuit' path and give a misleading reading. However, research at the University of Liverpool has shown that if measurements are taken at right angles to a single reinforcing bar the error is minimized as shown in Figure 4.14.

Obviously corrosion is an electrochemical process with current in the form of ions flowing through the concrete. The resistivity can tell us the capacity of the concrete to allow corrosion. It will not tell us if corrosion has started or if that capacity is being used to the full. Hence the earlier statement that at less than 10 kΩ cm resistivity is not the controlling parameter, but at more than 100 kΩ cm you cannot distinguish between active and passive steel as the resistivity will effectively stop corrosion.

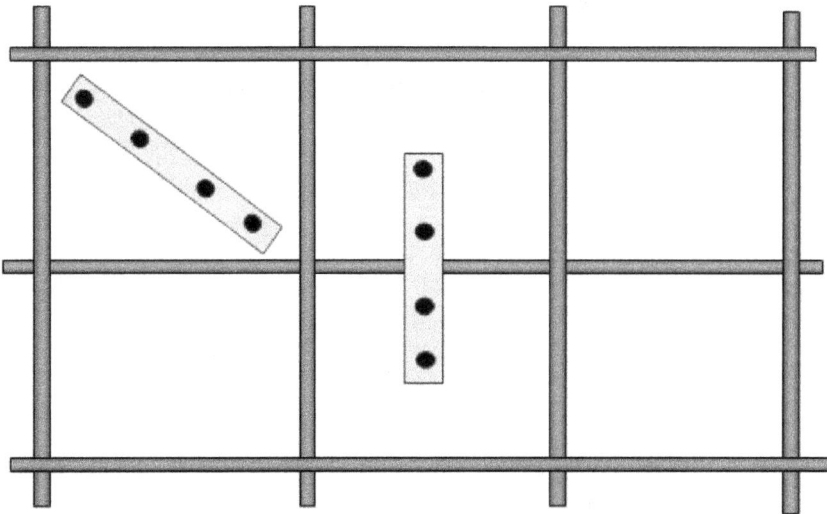

Figure 4.13 Minimizing the effect of reinforcing steel on four probe resistivity measurements when the probe array is bigger than the reinforcement spacing.

The resistivity calculation assumes the concrete to be homogeneous. The local inhomogeneity in electrical resistance of the aggregate must be allowed for by suitable probe spacing. The systematic inhomogeneity of the reinforcement network must be allowed for by minimizing its effect as shown in Figure 4.13. The effect of layers of different resistivity (e.g. the carbonated concrete zone or water penetration) is discussed and an 'influence diagram' given to show the effects of layers of different resistance on the measurement in Millard (1991).

4.11.5 Standards and guidance

There is no current standard on resistivity measurements other than British Standard BS 1881 Testing Concrete: Part 201: 1986 'Guide to the use of nondestructive methods of test for hardened concrete.' Concrete Society (2004) provides a detailed procedure and a pro forma for recording resistivity measurements. Concrete Bridge Development Group (2002) also gives information and guidance. There is also a RILEM Guide, Polder (2000).

4.12   Corrosion rate measurement

The corrosion rate is probably the nearest the engineer will get with currently available technology to measuring the rate of deterioration. There are various ways of measuring the rate of corrosion, including AC impedance and electrochemical noise (Dawson, 1983). However, these techniques are not suitable for use in the field for application to the corrosion of steel in concrete so this section will concentrate on linear polarization, also known as polarization resistance or LPR, and will discuss various macrocell or galvanic current measurement techniques.

Unfortunately we cannot yet use corrosion rate measurements to accurately calculate total section loss or predict concrete spalling rates but we can tell the engineer how much steel is turning into rust and how much metal is being lost at the time of measurement. The increasing use of corrosion monitoring systems (discussed in Chapter 5) will improve our ability to integrate the instantaneous rate to a loss of steel. The use of corrosion models to try to predict deterioration rates using chloride profiles, carbonation depths and corrosion rate measurements is discussed in Chapter 9.

4.12.1   Property to be measured

It is possible, with varying degrees of accuracy, to measure the amount of steel dissolving and forming oxide (rust). This is done directly as a measurement of the electric current generated by the anodic reaction:

$$Fe \rightarrow Fe^{2+} + 2e^-$$

and consumed by the cathodic reaction:

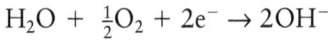

$$H_2O + \tfrac{1}{2}O_2 + 2e^- \rightarrow 2OH^-$$

and then converting the current flow by Faraday's law to metal loss:

$$m = \frac{Mit}{zF}$$

where m is the mass of steel consumed, i is the current (amperes), t is time in seconds, z is the ionic charge (2 for Fe $\rightarrow$ Fe^{2+} + 2e$^-$) and M is the atomic mass of metal (56 g for Fe). This gives a conversion of 1 μA cm^{-2} = 11.6 μm steel section loss per year.

4.12.2   Equipment and use – linear polarization

The linear polarization technique (also known as polarization resistance or conflated to LPR) requires us to polarize the steel with an electric current and monitor its effect on the reference electrode potential. It is carried out with a sophisticated development of the reference electrode incorporating an auxiliary electrode and a variable low voltage DC power supply. The reference electrode potential is measured and then a small current is passed from the auxiliary electrode to the reinforcement. The change in the reference electrode potential is simply related to the corrosion current (Stern and Geary, 1957) by the equation:

$$I_{corr} = \frac{B}{R_p} \tag{4.1}$$

where B is a constant (in concrete 26 to 52 mV depending upon the passivity or active condition of the steel surface) and R_p is the polarization resistance (in ohms)

$$R_p = \frac{\text{(change in potential)}}{\text{(applied current)}} \tag{4.2}$$

R_p gives the technique its alternative name of 'polarization resistance'. The change in potential must be kept to less than 20 mV or so for the equation to be valid and remain linear (hence the name 'linear polarization'). Also, there is an 'iR drop' in the circuit. This is the voltage that exists because a current is flowing through concrete which has a finite electrical resistance. This is also referred to as the 'solution resistance'. This can be measured by switching off the current at some point during the measurement process so that the potential without the 'iR drop' is measured. The iR drop is discussed further in the section on cathodic protection.

The measurement is made in one of two ways. Either steadily fixed levels of current are applied and the potential monitored (galvanostatic), or the current increased to achieve one or more target potentials (potentiostatic). In both cases allowances for the iR drop (solution resistance) may be made. A plot of change in current vs. change in potential gives a gradient of the polarization resistance R_p (equation 4.2) to calculate the steel section loss rate.

The corrosion rate in μm/year is:

$$x = \frac{11 \times 10^6 \cdot B}{R_p \cdot A} \tag{4.3}$$

where A is the surface area of steel measured in cm^2. A schematic of a typical linear polarization device is shown in Figure 4.14. This consists of a reference electrode with an annular auxiliary round which passes the current to the steel. The electrode may be stainless steel with conductive or wetted conventional foam to give a good contact to the concrete surface.

Defining the area of measurement is important for accurate corrosion rate measurement. In other (non-reinforced concrete) applications linear polarization is carried out on a sample or coupon of known size in or on, for instance, a pipe or tank or condenser where corrosion is a risk. However, we cannot put a sample into concrete, we must use the steel that is there which is all connected together and is in the corrosion environment, not some grout used to fix a coupon into the concrete.

The area of steel sensed by the electrodes is obviously far smaller than the whole rebar network, however we cannot assume that the area measured is that directly below the electrode as the current 'fans out' as shown in Figure 4.14. Assuming a 1 : 1 relationship between probe size and area of steel polarized can lead to errors of up to 100× in our measurement (Fliz et al., 1992). However, this is mainly at low corrosion rates. One device uses a large electrode with a conductive foam pad as the auxiliary electrode to minimize wetting of the concrete and ensure that the area of measurement is directly below the electrode as shown in Figure 4.15.

In some commercial devices a guard ring system has been developed to confine the area of the impressed current and thus define the parameter A in equation (4.3) which allows us to calculate the metal loss. One system uses extra reference electrodes to define the amount of current applied to the guard ring to ensure confinement of the signal. This is shown in Figure 4.17. Some devices have guard rings that apply fixed levels of current based on the auxiliary electrode current. These approaches are reasonably accurate at high corrosion rates but less accurate when corrosion rates

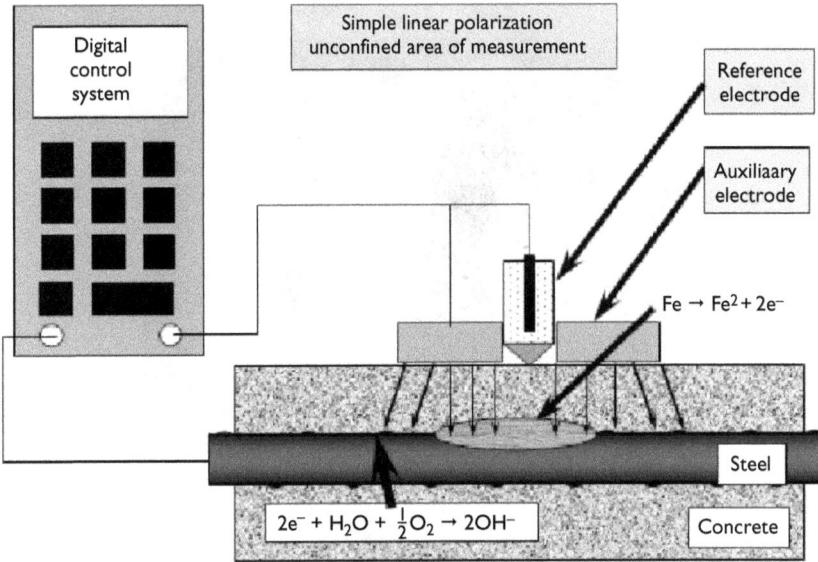

Figure 4.14 Schematic of a simple LPR system with unconfined measurement area.

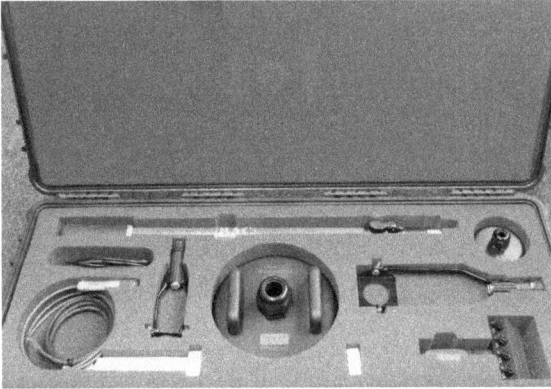

Figure 4.15 The BAC LPR Kit with 4 probe resistivity head (see Section 4.11).
Courtesy BAC Corrosion Control.

are low or localized (which may be when you may require greatest accuracy).

Criteria relating LPR readings to deterioration rates similar to the ASTM C876 criteria for reference electrode potentials have been published (Broomfield et al., 1993). These show some comparability between different devices and will be discussed later under interpretation (Section 4.12.4). A set of conversion equations is provided in the final report of the Strategic

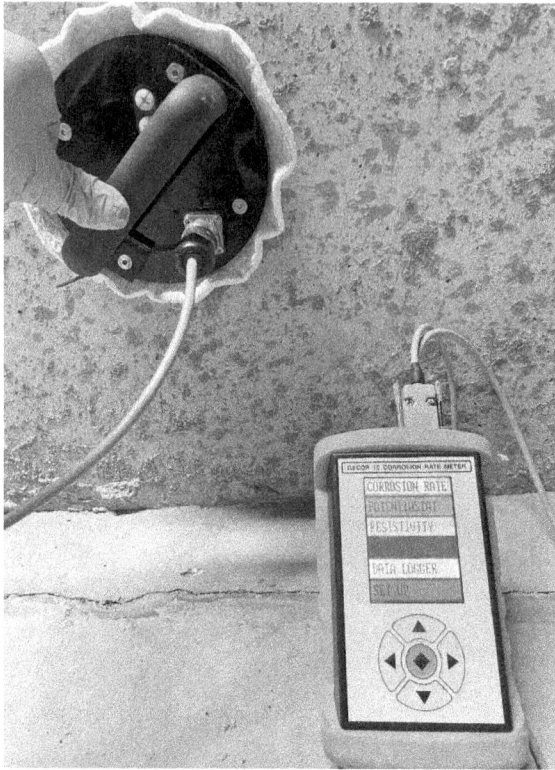

Figure 4.16 GECOR 10, a polarization resistance probe with reference electrode and 4 probe resistivity head (not shown). Courtesy SAFECOR Ingenieria SL.

Highway Research Program (SHRP) contract on corrosion rate measurement (Fliz *et al.*, 1992).

4.12.3 Carrying out a corrosion rate survey

Corrosion rate measurement is slow compared with reference electrode potential measurement. This is because the concrete reacts slowly to the electric field and changes must be reasonably slow to ensure that the electrochemistry in the concrete is changing linearly and without capacitance effects. However, the speed of the total operation varies significantly from device to device. The slowest devices are manually operated and take 10–20 min to take a reading that then must be manually plotted to calculate the corrosion current. These devices are rarely used today. The fastest microprocessor controlled devices take a few minutes and give the corrosion current directly. However, to be reasonably accurate, all devices must be set up over the reinforcement of known lay out and dimensions so that set up time is significant.

As the technique is slower than taking reference electrode measurements it is important to take measurements at the most significant locations on the structure, for example, by following up a potential survey with strategic corrosion rate measurements. Rate measurements should be taken at the positions of the highest and lowest potentials and at the steepest potential gradients.

It should also be noted that corrosion rates vary with the weather conditions. Corrosion rates accelerate in warm conditions. Resistivity will decrease as the concrete gets wet, also allowing corrosion rates to increase. For a full picture of corrosion conditions, measurements should be taken at regular intervals either throughout the year, or the same time each year so that results are comparable.

The corrosion rate is measured over a known area of rebar. This means that the rebars must be located and their sizes known so that the area of steel below the sensor is known. For purely comparative readings it can be adequate to take readings in comparable locations (such as isolated sections of bars or a cross over).

The potential must be stable throughout the reading so that a true change in potential (equations (4.1) and (4.2)) is recorded. This can lead to problems on very dry structures and where conditions are changing rapidly. Local wetting and coming back to the problem location later can sometimes overcome these problems.

Gowers *et al.* (1992) have used the linear polarization technique with embedded probes (reference electrode and a simple counter electrode) to monitor the corrosion of marine concrete structures. This technique was described previously without reference to isolating the section of bar to be measured (Langford and Broomfield, 1987). By repeating the measurement in the same location on an isolated section of steel of known surface area the corrosion rate of the actual rebar can be inferred. The main problem is the long-term durability of electrical connections in marine conditions. Also, the probes should ideally be built into the structure during construction. A desirable but rare occurrence. Recent developments in corrosion monitoring are described in Chapter 5.

A methodology and pro formas for recording results are given in Concrete Society (2004).

4.12.4 Interpretation – linear polarization

The following broad criteria for corrosion have been developed from field and laboratory investigations with the sensor controlled guard ring device shown in Figure 4.16 (Broomfield *et al.*, 1993, 1994):

$I_{corr} < 0.1 \ \mu A \cdot cm^{-2}$ Passive condition
$I_{corr} \ 0.1–0.5 \ \mu A \cdot cm^{-2}$ Low to moderate corrosion
$I_{corr} \ 0.5–1 \ \mu A \cdot cm^{-2}$ Moderate to high corrosion
$I_{corr} > 1 \ \mu A \cdot cm^{-2}$ High corrosion rate.

The device without sensor control has the following recommended interpretation (Clear, 1989):

$I_{corr} < 0.2 \ \mu A \cdot cm^{-2}$ No corrosion expected
$I_{corr} \ 0.2 - 1.0 \ \mu A \cdot cm^{-2}$ Corrosion possible in 10 to 15 years
$I_{corr} \ 1.0 - 10 \ \mu A \cdot cm^{-2}$ Corrosion expected in 2 to 10 years
$I_{corr} > 10 \ \mu A \cdot cm^{-2}$ Corrosion expected in 2 years or less.

These measurements are affected by temperature and relative humidity (RH), so the conditions of measurement will affect the interpretation of the limits defined earlier. The measurements should be considered accurate to within a factor of 2.

It should be noted that different values of 'B' are used by the two devices (equation (4.1)), with 26 mV used by the guard ring device and 52 mV used by the device without a guard ring. This may explain the factor of 2 difference in interpretation at the low end. At the high end, the lack of a guard ring may lead the simpler device to include a larger area of steel in its measurement, indicating a higher corrosion rate. Alternatively, the device may have been used on more actively corroding structures and the interpretation range may therefore have been extended. An equation linking the devices and a third, simple guard ring, is given in Fliz et al. (1992). The correlation factors (R^2) showed relative accuracies of the devices of 70–90%.

4.12.4.1 Effects of temperature

Temperature affects the corrosion rate directly. The rate of the oxidation reaction is affected by the amount of heat energy available to drive the reaction. Also the concrete resistivity reduces with increased temperature as ions become more mobile and salts become more soluble. It also affects the RH in the concrete, lowering it when temperature increases and therefore introducing an opposite effect. At the other extreme the pore water will freeze and corrosion stops as the ions can no longer move. It should be noted that this will happen well below ambient freezing point as the ions in the pore water depress the freezing point well below 0°C. This is shown in Figure 4.17 for carbonated concrete. In a paper on the results from a permanent corrosion monitoring system in a highway tunnel, Broomfield et al. (2003) found corrosion rates fluctuated by a factor of 2.5 when the tem-

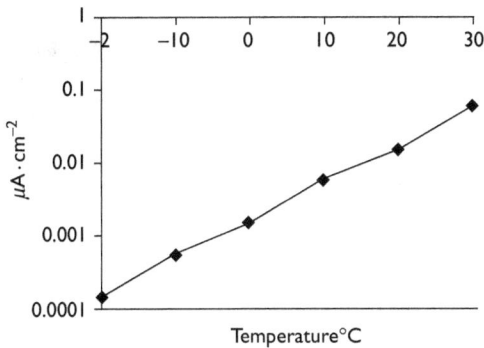

Figure 4.17 The effect of temperature on carbonated concrete (based on Tuutti, 1982).

perature varied from 2 to 25°C. There was no significant corrosion but the rates ranged from 0.08 to 2 μm/year.

4.12.4.2   Effects of RH

This will also be a factor in determining how much water is there in the pores to enable the corrosion reaction to be sustained. Chloride-induced corrosion is believed to be at a maximum when the RH within the concrete is around 90–95% (Tuutti, 1982). This is shown in Figure 4.18 for specimens exposed to a 3% NaCl solution (approximation of sea water concentration). For carbonation there is experimental evidence from the same source that the peak is around 95–100% RH (Figure 4.19). However, it is important to recognize that the RH in the pores is not simply related to the atmospheric RH, water splash, run off or capillary action, formation of dew, solar heat gain, thermal mass of the structure and other factors mean that atmospheric RH is often quite different from that seen by the reinforcing steel.

Increased water saturation will slow corrosion through oxygen starvation once the pores are filled with water and oxygen cannot get in. Conversely, totally dry concrete cannot corrode due to the absence of water. However, highly water saturated structures can corrode rapidly without signs of cracking. This is due to the limited amount of oxygen available as the iron ions (Fe^{2+}) stay in solution without forming solid rust that expands and cracks the concrete. A totally saturated structure can reach a very high negative reference electrode potential (-900 mV) without corroding due to oxygen starvation. However, if oxygen then does get access to the steel the corrosion rate will be very high as the steel will have no passive oxide layer to protect it.

This is one of the reasons why corrosion rate measurement can be very important in finding the true situation in such ambiguous conditions.

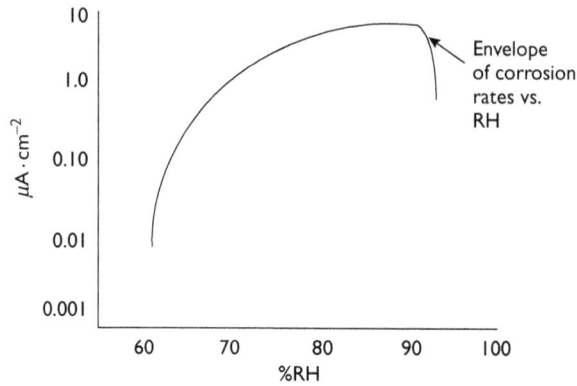

Figure 4.18 Variation in corrosion rate with RH in chloride contaminated concrete (based on Tuutti, 1982).

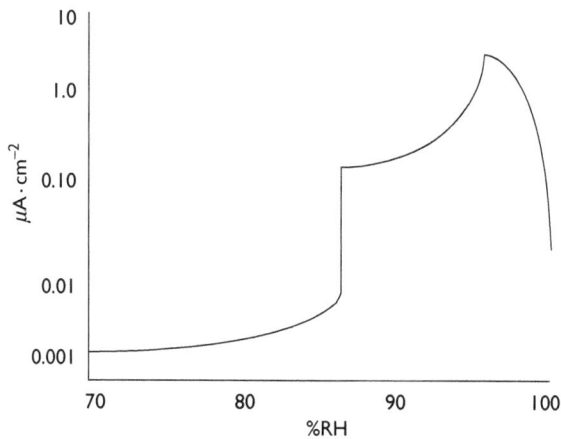

Figure 4.19 Variation in corrosion rate with RH in carbonated concrete (based on Tuutti, 1982).

4.12.4.3   Pitting and rapid corrosion

Much of the research on linear polarization has been done on highway bridge decks in the United States where chloride levels are high and fairly uniform. Bridge decks are rarely protected with membranes and overlays and therefore pitting is rarely observed (see Section 2.3). However, in Europe, where corrosion is localized in run down areas on bridge substructures and pitting is more common, the problems of interpretation can be complicated as the I_{corr} reading is coming from isolated pits rather than uniformly from the area of measurement.

Figure 4.20 A post-tensioned marine pile showing rapid tendon section loss after removal of a jacketing repair from Rapa, (1999). Courtesy Florida Department of Transportation.

Laboratory tests with the guard ring device have shown that the corrosion rate in pits can be up to 10 times higher than generalized corrosion. This means that the device is very sensitive to pits. However, it cannot differentiate between pitting and generalized corrosion. That must be done by direct observation of the steel or by careful study of the reference electrode potentials and chloride contents. In a long-term study of the corrosion of concrete posts the author observed that corrosion rates increased with time, then decreased just before cracking and spalling was observed on the structures. One tentative conclusion was that the rise in corrosion rate measurement was associated with pit formation where high corrosion rates are observed. As the pits widened and joined up the corrosion rate decreased but the oxide formation was more generalized and cracking ensued (Broomfield *et al.*, 2000).

In the photograph in Figure 2.3 we can see a plane reinforcing bar approximately 25 mm in diameter, reduced in section to about 8 mm. The bar was from a highway bridge built in the 1960s or 1970s. The bar was extracted in the 1990s. Therefore in less than 30 years it had lost 17 mm of its section, that is, a rate of more than 0.5 mm/year (this assumes that corrosion started as soon as the bridge was built). In this case corrosion was caused by deicing salt penetrating the concrete and attacking the steel. We know that it takes several years for sufficient deicing salt to penetrate concrete to the reinforcement and start corrosion. It is therefore probable that corrosion was occurring at 1 mm/year or faster. In this case the corrosion was on steel in good quality modern concrete but caused by deicing salt ingress.

In another case (Figure 4.20) prestressing strands in a marine pile were found to have corroded through shortly (three years) after repair. The tendons were 7/16 inch in diameter. This implies a rate of over 3 mm/year if a 7/16″ = 11 mm tendon loses its full section in three years. This demonstrates that corrosion rates in concrete can be very high and have nothing to do with generalized corrosion rates for atmospheric conditions.

While these examples pertain to steel suffering from chloride-induced corrosion, the author has also seen severe section loss on steel which had no chlorides. However, this was a clinker concrete. These are known to be acidic when wet and very aggressive to steel and are no longer used for reinforced concrete (Neville, 1995).

4.12.4.4 Predicting deterioration rates

Measurements can be used to predict the deterioration rate of a structure when used in combination with diffusion data, chloride or carbonation penetration, and cover measurements. The ongoing rate of deterioration may be important in deciding whether to defer rehabilitation work, carry out minor cosmetic repairs in anticipation of later major works, or whether to bring forward major works. It may also be used to prioritize repairs either on an element by element or a structure by structure basis. This is further discussed in Chapter 9.

Some work has been done in translating I_{corr} to loss of cross section of reinforcement and hence end of service life (see equation 4.3 and Andrade *et al.*, 1990). However, the loss (spalling) of concrete is the most usual cause for concern, rather than loss of reinforcement strength. It is far more difficult to predict cracking and spalling rates especially from an instantaneous measurement. A simple extrapolation, assuming that the instantaneous corrosion rate on a certain day is the average rate throughout the life of the structure, often gives inaccurate results (Broomfield *et al.*, 1993). This is true for section loss. Converting that to delamination rates is even less accurate as it requires further assumptions about oxide volume and stresses required for cracking.

One approach is to work on the estimates of how much section loss will generate enough expansive oxide growth to cause cracking. This has been found to be between 10 and 100 μm (0.01–0.1 mm).

The oxides produced by iron have volume increases between twice and six times the steel consumed as shown in Figure 2.2. This assumes that they are 100% dense (with no porosity occupying extra space). To generate cracking the oxides must be deposited at the metal/oxide interface.

If we assume that the average expansion ratio is 3, allowing for oxide depositing elsewhere and some porosity in the oxide, then the corrosion rates given earlier translate as follows:

0.1 $\mu A \cdot cm^{-2}$ = 1.1 μm/year section loss = 3.3 μm/year rust growth
0.5 $\mu A \cdot cm^{-2}$ = 5.5 μm/year section loss = 16.5 μm/year rust growth
1.0 $\mu A \cdot cm^{-2}$ = 11.5 μm/year section loss = 34.5 μm/year rust growth
5.0 $\mu A \cdot cm^{-2}$ = 57.5 μm/year section loss = 173 μm/year rust growth
10 $\mu A \cdot cm^{-2}$ = 115 μm/year section loss = 345 μm/year rust growth.

In laboratory tests, Rodríguez (1994) found 15–40 μm section loss gave rise to cracking on bars with a cover/diameter ratio between 2 and 4. This is around the 0.5 $\mu A \cdot cm^{-2}$ transition from low to moderate corrosion rate.

4.12.4.5 Before and after measurements

Another major application is in defining the effectiveness of a repair. The application of a silane, an inhibitor or of chloride removal or realkalization can lead to the need for long-term monitoring to see if the treatment reduces the corrosion rate in a before and after comparison. Long-term monitoring is recommended to ensure that the treatment is still working.

4.12.5 Limitations – linear polarization

There are two major limitations to the linear polarization devices described earlier:

1   They detect the instantaneous corrosion rate which can change with temperature, RH and other factors.
2   Either they make assumptions about the area of measurement or they define an area of measurement.

The first factor is a limitation because we do not want to know the instantaneous corrosion rate. We actually want to know the integrated corrosion rate from the time of initiation of corrosion. We must therefore make assumptions or multiple measurements with time to estimate the average or integrated rate.

The second limitation is that there can be errors of 10–100 in the estimated area of measurement especially at low corrosion rates, unless the sensor controlled system is used. Even if the sensor controlled system is used it gives a single value for the steel contained within a 105 mm circle below the sensor. If corrosion is at a few pits it will underestimate the corrosion rate in the pits and overestimate the general corrosion rate. If it is known that there is pitting corrosion then corrections can be made based on the fact that corrosion rates in pits are 5–10 times that of generalized corrosion.

There are further errors because the value of B in equation (4.1) varies from 26 to 52 mV depending upon whether the steel is active or passive. Further, corrosion may be concentrated on the top of the bar, or, if bars are close together or deep within the concrete, the device may only send current to the top steel. Both of these errors mean that the best accuracy you can expect from a linear polarization device is a factor of 2 to 4 (Andrade *et al.*, 1995). This is supported by the discussion in Section 4.12.4 which showed that seasonal fluctuations give a factor of 2.5 variation in LPR measurements. However, the scale is logarithmic so such errors are less critical than they seem to be.

The final limitation is the difficulty of converting the corrosion rate in terms of rate of metal section loss to a total section loss or cracking and spalling rate. Further work in this area is leading to the resolution of this problem.

4.12.6 Equipment and use – macrocell techniques

An alternative approach, which reintroduces the theme of long-term corrosion monitoring, is the embedding of macrocell devices. This includes galvanic couples of different steels (Beeby, 1985) or embedding steel in high chloride concrete to create a corrosion cell, as is popular in cathodic protection monitoring systems, particularly in North America (NACE, 2000).

This approach is also popular in laboratory corrosion studies and has been developed as an ASTM procedure known as ASTM G109 (2021) in which concrete prisms are made with a single top rebar in chloride containing concrete and two bottom rebars in chloride-free concrete. The current flow between the top and bottom is monitored as shown in Figure 4.21.

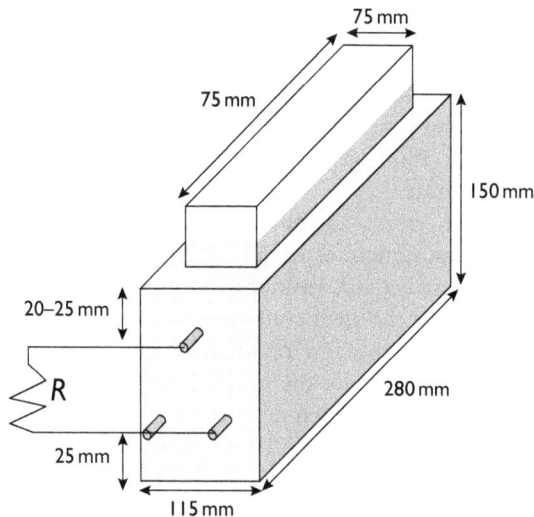

Figure 4.21 Schematic of ASTM G109 macrocell current prism.

Another development has been used for installing in new structures, for instance the Great Belt Bridge in Denmark (Shiessl, 1993). This uses a 'ladder' of steel specimens of known size, installed at an angle through the concrete cover. As the chloride (or carbonation) front advances each specimen becomes active and the current flow from the anodic specimen to its adjacent cathodic specimen can be monitored along with the reference electrode potential. This system is discussed fully in Chapter 5.

4.12.7   Interpretation – macrocell techniques

There are two major problems with interpreting macrocell systems. The first is that we are not actually measuring corrosion on the reinforcing steel. We must assume that the current flows are representative of what is happening on the steel we want information about. That will partly depend upon the care and thought taken in the installation of the probes.

The second issue is how representative are macrocell currents of the true corrosion currents in the steel. The microcell currents may be more important than macrocell current flows. In a comparison with linear polarization (Berke *et al.*, 1990) it was found that the macrocell technique underestimates the corrosion rate, sometimes by an order of magnitude. As this was using the ASTM prism technique, it should be considered the most accurate use of the macrocell technique, so if it is an order of magnitude out, field use of macrocell techniques could be even less accurate.

The advantage of these techniques is that they are permanently set up and can be used to monitor the total charge passing with time. This is a clearer indication of the total metal loss or total oxide produced than an instantaneous measure of the corrosion rate. Also, the measurement equipment is fairly straightforward, being a voltage measurement across a resistor or a direct zero (or low) resistance ammeter measurement of the current.

4.12.8   Standards and guidance

The macrocell test ASTM G109 (2021) is the only standard for measuring macrocell corrosion rates of steel in concrete. Guidance on LPR comes from Concrete (2000) and from the manuals provided by manufacturers and suppliers of the equipment.

4.13   Permeability and absorption tests

Since corrosion of steel in concrete is usually caused by the ingress of various agents (Cl^-, CO_2, H_2O, O_2) through the concrete cover, many attempts have been made to calculate and measure the permeability and absorbtion characteristics of the concrete. Measurements are most accurately done on a conditioned core in the laboratory, since the environment (i.e. degree of saturation) will strongly influence field measurements. However, field measurements are made with vacuum devices and water absorption kits to give a rating of the permeability.

4.13.1 Property to be measured

In the laboratory, diffusion, permeability permittivity and absorption can all be measured under controlled conditions. When conducting such an investigation it is important to differentiate between permeability which is measured under steady state conditions driven by a constant pressure difference, and initial surface absorption driven by rapid capillary absorption (Concrete Society, 1997a, p. 54).

4.13.2 Equipment and use

Most laboratory tests are concerned with modelling the flow of chloride into concrete. Unfortunately even the most accurate measurements in the laboratory can rarely be equated to field conditions where wetting and drying accelerate chloride uptake and the surface concentration is rarely either known or constant so steady state diffusion is a poor model at best.

'Effective' chloride diffusion coefficients can be back calculated from the fitting of chloride profiles to diffusion curves using the error function equation. An alternative technique is to plot the square root of the effective concentration (total minus background) vs. depth, and extrapolating a straight line fit to the Y axis. This method is discussed in Chapters 3 and 9.

Field devices are most effective when used to check improvements in water absorption after a treatment such as the application of a penetrating sealer such as silane. The ISAT test (BS 1881-208, 1996) is particularly useful for checking the effectiveness of coatings.

4.13.3 Interpretation

Concrete Society Technical Report 50 is a guide to surface treatments for protection and enhancement of concrete (Concrete Society, 1997a). This gives the following criteria for the ISAT test:

- Good concrete judged to have low absorption <750 g/m^2·h$^{0.5}$
- Most coatings can achieve <35 g/m^2·h$^{0.5}$
- Very impermeable coatings can achieve <10 g/m^2·h$^{0.5}$.

A European standard BS EN 1062-3 (2008) has been published. Vacuum tests which equate air movement with moisture permeability and other devices give approximate indications of the rate of diffusion (Whiting *et al.*, 1992; Concrete Bridge Development Group, 2002).

4.13.4 Limitations

Most laboratory tests are concerned with modelling the flow of chloride into concrete. Unfortunately even the most accurate measurements in the laboratory can rarely be equated to field conditions where wetting and drying accelerate chloride uptake and the surface concentration is rarely either known or constant so steady state diffusion is a poor model at best.

'Effective' surface concentrations are used by some researchers to back calculate diffusion constants from the measured concentrations. These are usually the concentrations measured at 5–13 mm depth, where the chloride concentration is less affected than at the the the surface where recent washing, drying, etc. may deplete or enhance the chloride concentration (see Figure 3.2(b)).

An alternative technique is to plot the square root of the effective concentration (total minus background) vs. depth, and extrapolate a straight line fit to the Y axis. This method is discussed in Chapters 3 and 9.

4.13.5 Standards and guidance

The ISAT test (BS 1881-208 1996) is particularly useful for checking the effectiveness of coatings. Concrete Society Technical Report 50 is a guide to surface treatments for protection and enhancement of concrete (Concrete Society, 1997a). Rilem Report 12 (Kropp and Hilsdorf, 1995) goes into great detail on the various tests and transport phenomena in concrete.

4.14 Concrete characteristics: cement content, petrography, W/C ratio

It is usually essential to take cores and carry out selective petrographic analysis during the full survey. This can be used to determine the types of aggregates (susceptibility to ASR, freeze thaw damage, sulphate attack, presence of highly absorptive constituents, etc.), mix design (cement, water and aggregate ratios) and the extent of curing and hydration.

It is normal (in the United Kingdom and Europe) to determine the chloride concentration as a percentage by mass of sample but to quote chloride concentrations in weight percent of cement so the cement content must be known or estimated (usually assumed to be 14%). This is not a very accurate procedure which is why in North America they quote chloride contents in pounds per cubic yard of concrete. The European approach has a better grounding in the mechanism of corrosion, while the North American approach is more accurate in practice, but slightly less informative.

Standards for petrographic analysis include: ASTM C457-98 – Standard test method for microscopical determination of parameters of the air-void system in hardened concrete; ASTM C856-95e1 – Standard practice for petrographic examination of hardened concrete; BS 1881 Testing concrete – Part 124: 1988 Methods of analysis of hardened concrete; and BS812-104: 1994 – Testing aggregates. Methods for quantitative and qualitative petrographic examination of aggregates.

4.15 Ground penetrating radar

Ground penetrating radar (GPR) is an electromagnetic technique that has been used for many years within the engineering industry to supply information on construction and condition detail of various types of structure. GPR, also known as impulse radar, gives valuable information on the location of reinforcement and can often supply information where cover meters cannot, such as lapping bars, multiple layers and tendon ducts behind cover reinforcement. There has already been some discussion of its application for delamination measurement in Section 4.6.

4.15.1 Property to be measured

The GPR is an electromagnetic echo sounding method where a transducer (transmitter/receiver) is passed over the surface at a controlled speed. Short duration pulses of radio energy are transmitted into the subject and reflections from material boundaries and embedded features such as metalwork or voids are detected by the receiver. Sampling is so rapid that the collected data are effectively a continuous cross section, enabling rapid assessment of thickness and condition over large areas. By assessing the strength, phase and the scatter of signals it is often possible to find cracking and changes in compaction, bond and moisture content. However, the main application is to locate reinforcement and tendon ducts (arrangement and depth of cover) as well as delaminations, voids, cracks and honeycombing. Section 4.6 discusses the use of vehicle mounted radar to assess delaminations on bridge decks with vehicle mounted equipment.

4.15.2 Equipment and use

The GPR equipment is generally portable and self-contained enabling access to most structures. The antenna can be operated from the end of a long cable enabling large areas to be covered relatively easily. The technique demands hand access to the structure so hoists or scaffolding are often required. In North America vehicle mounted systems are used, as described in Section 4.4, on concrete and asphalt-covered bridge decks.

When assessing reinforced concrete structures it is recommended that high frequency antenna are used, where possible, in order to give the highest resolution image. Antenna of 1.5 GHz are most commonly used to give the required amount of information and will give useful information typically to a depth of *c*.300 mm. Where greater penetration is required a lower frequency antenna should be used. Recent developments in higher frequency antennas, up to 4 GHz, may result in much higher resolution in the future.

In order to accurately relocate any features identified from the investigation it is recommended that an electronic distance measuring (EDM) wheel is linked to the GPR system. This enables scans to be taken at fixed distance intervals and removes the necessity for surveying at a constant speed that is required from time triggered systems. Typically scans would be taken at *c*.5–10mm intervals to give the required lateral resolution.

It is essential that all collected data are recorded digitally in order that post-processing and quality control can take place. Filtering of data off site can often result in higher levels of quality as well as reducing errors from digitizing of analogue plots.

4.15.3 Interpretation

Data collection is far more rapid than using cover meters or other hand-held instruments. However, interpretation is far more complex as many features in the concrete will reflect the radar waves. A radar trace is shown in Figure 4.22 showing honeycombing and reinforcing bars.

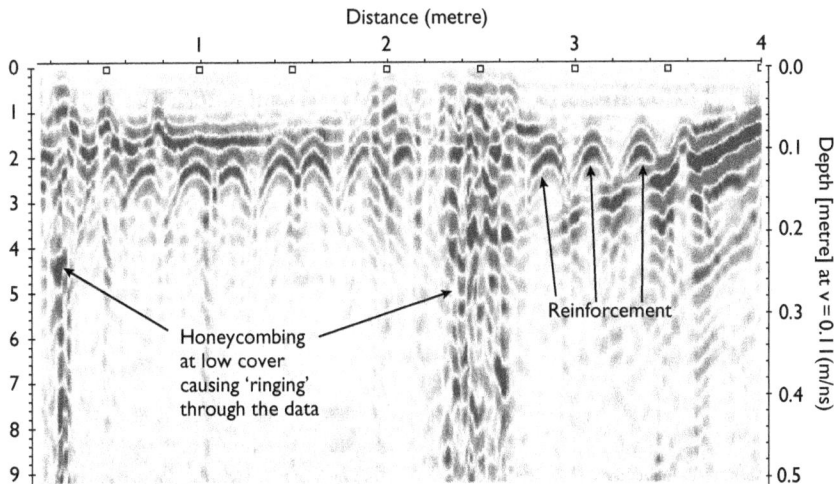

Figure 4.22 Radar trace showing honeycombing and reinforcing bars. Courtesy Fugro Geotechnics Ltd.

4.15.4 Limitations

Vehicle mounted radar has been found to be reasonably accurate in predicting the amount of damage on a bridge deck, but not the precise location of the damage. The problem for both vehicle mounted and hand-held systems is the number of features in the concrete that will cause reflections. It can therefore be difficult to resolve types of features (delaminations from honeycombing from patch repairs). It is also difficult to resolve features at different depths. In the United Kingdom and the United States, GPR services are offered by specialist companies who interpret the data using proprietary algorithms as well as their extensive experience in conducting radar surveys of reinforced concrete.

4.15.5 Standards and guidance

ASTM D4788-03 (2003) gives a standard test method for its application to bridge decks, which can be used as guidance for its application on other concrete elements.

4.16 Ultrasonic pulse velocity

The pulse velocity approach is best done in a 'transmission mode' with the pulse created on one side of a concrete member and detected on the other side. It can be used at corners or in a 'reflective' mode if necessary, but it loses effectiveness and interpretation gets harder. The impact-echo technique can be used with pulse generator and detector side by side as the echo is reflected back from defects.

The UPV involves the measurement of acoustic wave velocity and gets the name ultrasonic from the range of frequencies beyond the audible range that it employs: $c.20 - c.300$ kHz. In order to determine the wave velocity a transmitter and receiver are placed onto the surface of the structure at a measured distance apart. The time taken for a pulse to travel from the transmitter to the receiver is measured. Good contact generally requires the use of a coupling gel between the transducers and the structure is necessary when using the PUNDIT (Portable Non-Destructive Digital Instrument Tester).

4.16.1 Property to be measured

Estimating concrete strength from ultrasonic measurements using a PUNDIT is a well tested technique. Recent developments have meant that the technique can also be used in reflection mode as an assessment of condition as well as in the more traditional transmission mode for strength estimation.

4.16.2 Equipment and use

The Pundit Plus for use in transmission mode is shown in Figure 4.23. The A1220 Ultrasonic Flaw Tester is an example of a multi-headed ultrasonic probe for use in reflection mode as shown in Figure 4.24.

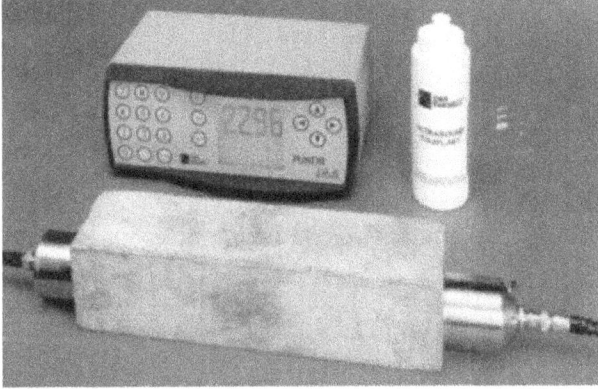

Figure 4.23 The Pundit Plus system for use in direct transmission mode.

Figure 4.24 The multi headed ultrasonic flaw detector for use in reflection mode. Courtesy Fugro Geotechnics Ltd.

4.16.3   Interpretation

The analysis of direct travel times can be used to:

- relate velocity to the mechanical properties of the materials such as the elastic moduli and the uniaxial compressive strength;
- locate and identify defects such as cracks and voids;
- determine the depth of penetration of a crack (to a maximum depth of approximately 250 mm);
- determine the depth of fire damaged concrete.

In the reflection mode the technique can be used to:

- determine thickness of concrete;
- locate voiding, delamination within the concrete.

Figure 4.25 shows the combination of data from radar and ultrasonic to resolve reinforcement location and delaminations in a reinforced concrete structure of unusual construction.

Figure 4.25 Combined GPR and ultrasonic investigation to resolve reinforcement location (diagonal and horizontal bars) as well as delaminations using a A1220 low frequency flaw detector in reflection mode. Courtesy Fugro Geotechnics Ltd.

4.16.4 Limitation

The collection of ultrasonic data is relatively slow due to the need to collect the data from point locations. The results often give a relative condition only and as such a good knowledge of the structure is necessary. As with many non-destructive tests it is often necessary to calibrate the findings from limited targeted intrusive work.

The surface of the structure to be investigated needs to be smooth in order to get sufficient energy into the concrete. Whereas gel couple methods such as the pundit require a smooth surface the dry couple transducers can accommodate a certain amount of roughness and therefore less surface preparation is necessary.

On many sites external noise sources may interfere with the instrument and this should be discussed to avoid wasted trips to site. Many factors such as traffic movement and general site noise should not affect the instrument as the frequency would be outside the recording range of the equipment. Needle guns are a source of interference which would affect the technique.

Although ultrasonics in reflection mode may give some information on presence of reinforcement it is not a suitable method for accurate assessment of reinforcement detail.

4.16.5 Standards and guidance

The use of ultrasonic pulse velocity is documented in BS EN12504-4:2000 'Testing Concrete. Determination of ultrasonic pulse velocity'.

4.17 Impact-echo

Impact-echo is a method for non-destructive evaluation of concrete and masonry structures, based on the use of impact-generated stress (sound) waves that propagate through a structure and are reflected by internal flaws and external surfaces.

4.17.1 Property to be measured

Impact-echo can provide the thickness of concrete slabs up to an accuracy of 3% and can determine the location and extent of flaws such as cracks, delaminations, voids, honeycombing and debonding in plain, reinforced and post-tensioned concrete structures. It can also locate voids in the subgrade beneath slabs and pavements. For masonry it can determine thickness and locate cracks, voids and other defects where the brick or block units are bonded together with mortar. Impact-echo is not adversely affected by the presence of steel reinforcing bars.

4.17.2 Equipment and use

The typical equipment is portable and battery powered, enabling it to be used in most locations. The equipment comprises a control unit and transducer with the system being run from a laptop computer. The energy source used consists of spherical impactors (ball bearings) which are used to strike the concrete surface to set up standing waves within the structure. The smaller the impactor the higher the frequency and therefore the higher resolution.

Figure 4.26(a) shows typical equipment used for impact-echo measurement and interpretation

4.17.3 Interpretation

If the speed of sound through the concrete is known or can be determined on site then it is possible to determine the depth of features immediately as in Figure 4.26(b) where the time domain and frequency domain plots are shown.

4.17.4 Limitations

As with several other techniques discussed in the last few sections, this technique requires specialized expertise and is therefore usually conducted by test houses who use it on a regular basis. Intimate contact is required between the impactor and the surface.

4.17.5 Standards and guidance

ASTM C597 'standard test method for pulse velocity through concrete' and ASTM C1383 'test method for measuring the P-wave speed and the thickness of concrete plates using the impact-echo method' cover the application of impact-echo to concrete structures.

4.18 Radiography

Radiography uses high frequency electromagnetic radiation (X-rays and gamma rays) to gain detailed images of the internal structure of concrete bodies. This method is more commonly known for its applications in the field of medical science. Despite the highly specialized equipment and safety precautions, radiography produces the most readily understandable results.

The principle behind radiography involves the measurement of attenuation of radiation as it passes through the material under investigation. The radiation is provided by a focussed source directed at one side of the structure which is detected /recorded on the other side by photographic film or a detection plate.

(a)

(b)

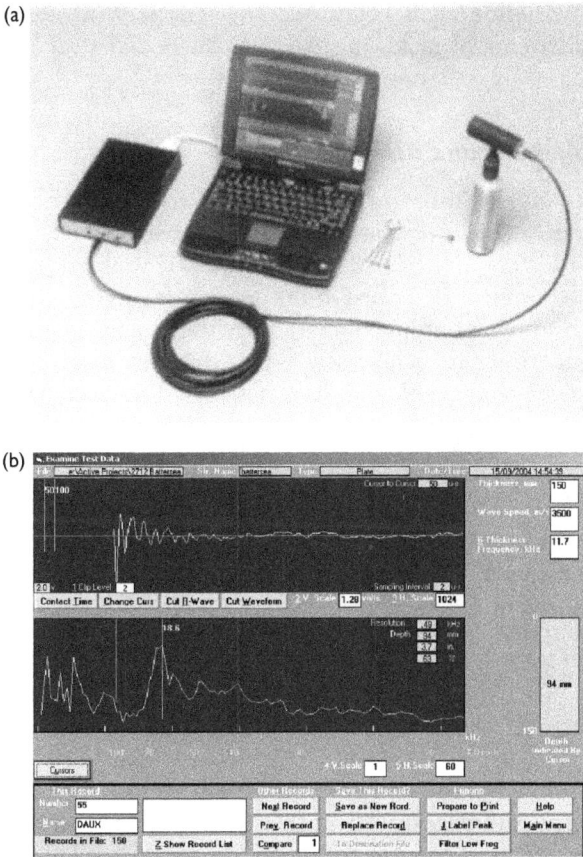

Figure 4.26 (a) Impactor/transducer, electronic interface and laptop computer for generating the impact and recording the echo spectrum. Courtesy Physical Acoustics Ltd. (b) Impact-echo response showing a delamination at 94 mm depth (lower frequency domain plot). Courtesy Fugro Geotechnics Ltd.

The resulting images can be produced relatively quickly to allow preliminary interpretation on site.

4.18.1 Property to be measured

As the radiation is transmitted through the concrete it interacts with the atoms in the material. This interaction causes some of the radiation to be absorbed by the concrete and some to be scattered. The result is that the amount of radiation detected varies spatially depending on the density and nature (atomic number) of the material(s) of the structure under investigation.

Therefore, variations in the concrete (such as cracks, voiding and compaction variation) will be caused by variations in density and the location and extents of inclusions (such as ducts and steel reinforcement) will be produced by variations in material types.

4.18.2 Equipment and use

A typical site investigation may use a portable X-ray system such as the Betatron (Figure 4.27). The system comprises three units with interconnecting cables with a total weight of 190 kg. Although this is bulky compared to the other equipment described in this section this system is compact and portable with a power output comparable with permanent systems (capable of penetrating concrete in excess of 1 m thick in ideal circumstances). Smaller systems are available that have associated reduced penetration and resolution.

4.18.3 Interpretation

A typical image is shown in Figure 4.28 before and after image clean up. Metallic items embedded in the concrete can be clearly seen, even down to some section loss of the reinforcement.

4.18.4 Limitations

For all systems the site considerations are the same:

- Access is required to both sides of the structure.
- The site needs to be screened and secure to provide a safe environment.
- Progress rates tend to be slow so time limitations and coverage need to be considered.

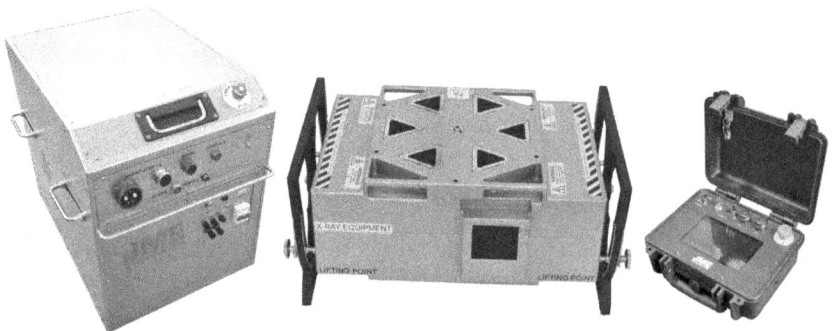

Figure 4.27 Portable 7 MeV X-ray Betatron. Courtesy of JME Ltd.

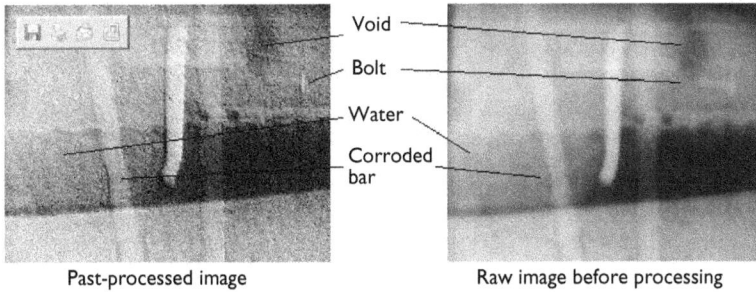

Void
Bolt
Water
Corroded bar

Past-processed image | Raw image before processing

Figure 4.28 Raw and processed images of a radiogram of a concrete structure. Courtesy of JME Ltd.

- Multiple layers of reinforcement will be superimposed and may limit interpretation.
- Although interpretation can be straightforward the specialized equipment and safety considerations require expert operators.

4.18.5 Standards and guidance

General information regarding radiography of concrete structures can be found in BS 1881: Part 205. 'Concrete Testing. Recommendations for radiography of concrete'. National regulations on ionizing radiation should also be consulted.

4.19 Survey and assessment methodology

When carrying out a survey (or commissioning a specialist test house) it is important to define what information is needed and how that information can be collected accurately but economically. If access is a problem (e.g. on a bridge or a working plant) it may be important to collect available information during a single access period rather than go through the expense or difficulty of providing repeat access. On highway structures access costs including traffic control can be the largest cost of the job.

Plan what measurements will be taken and where they will be taken. Allow for flexibility as the information changes during the survey. Try to check and analyse data from the survey as it proceeds. If potentials are high in one area that may be the best place for coring and drilling (as well as taking some representative samples from other areas).

Plan to spend significant resources on the survey. It could save you from wasting a lot of money later on. Have regard for the repair options available.

If an electrochemical technique is under consideration it may be worth checking for the degree of continuity of reinforcement as good continuity is essential but may be difficult or expensive to establish. Time and effort spent at this stage may produce later savings. Conversely a poorly executed survey can produce problems if a rehabilitation technique based on the survey results turns out to be uneconomic or impractical. Examples that the author has come across include expensive patch repairs and coatings that failed after less than a year leading to later installation of a cathodic protection system, doubling rehabilitation costs, and a cathodic protection system which had to be redesigned when it was found that there was two to five times as much steel in the structure as originally estimated.

An important aspect when considering cathodic protection or the other electrochemical techniques is zero cover or tie wire touching the surface. If there is a direct metallic connection between the anode and the rebar it could short out the electrochemical protection if a surface applied anode is under consideration. Excessive shorts have caused CP systems to be abandoned in the United States. On the other hand, severe congestion of the reinforcement can make the installation of discrete anodes difficult or uneconomic. Contrary to popular opinion, lack of electrical continuity between reinforcing steel bars is rarely a problem and is usually fairly straightforward to overcome. These issues are discussed further in Section 7.6 on cathodic protection.

When planning a survey it is important to know what information you want by the end of it. While executing the survey it is important to interpret the data as you go to ensure that the most useful measurements are taken and then new information is used to draw the correct conclusions, not just the expected ones. The survey report should draw conclusions and make recommendations, not just be a summary of the data.

Concrete Bridge Development Group (2002) has information on UK highway bridge contract documentation, specification and 2001 pricing for surveying UK highway bridge elements.

4.20   Special conditions – prestressing and coated rebars

There are particular problems when the reinforcing steel is not of the conventional type of steel bars embedded directly into the concrete. There are problems for galvanized steel bars with respect to reference electrode and corrosion rate measurement because the zinc affects the readings in poorly understood ways. However, when the bar is coated in epoxy or the reinforcement is in the form of wires in ducts the problems are multiplied as described later.

4.20.1   Epoxy coated and galvanized reinforcing bars

Epoxy coated rebars present particular problems for determining the corrosion condition of the steel. In the first place the bars are electrically isolated from the concrete except at areas of damage. The size and locations of the areas of damage are obviously unknown. Attempts to carry out reference electrode potential surveys and linear polarization measurements have therefore been unable to come up with definitive criteria for corroding and non-corroding areas. The other problem is that unless there have been problems during rebar cage fabrication the bars are isolated from each other, therefore a connection must be made to each bar measured to be sure that there is electrical contact.

When the steel is covered in other coatings, such as galvanizing, then the potentials are created by the zinc, not the steel. This leads to very negative potentials (about -500 mV to -800 mV with respect to a Ag/AgCl or calomel electrode) when the structure is new. This drifts down to less negative values either as the zinc passivates in the alkaline concrete or as it is consumed and the steel becomes active. It can be impossible to distinguish the two conditions so potential measurements are not useful on galvanized structures.

Corrosion rate measurements should be representative of galvanized steel but the 'B' value in equation (4.1) for zinc may be very different from that of steel in concrete. The author is not aware that any B values for zinc or galvanized steel in concrete have been published. However, comparative results may be useful in showing areas of high and low corrosion.

4.20.2 Internal prestressing cables in ducts

Pretensioned prestressing systems with wires or rods embedded directly in the concrete can be treated as though they are reinforcing bars, allowing for the different metallurgy. However, the problem of prestressing steel cables in internal post-tensioned structures with ducts is more difficult. These structures are usually built of precast elements that are either individually or collectively stressed by cables in ducts running deep within the structures (cover of typically 50–100 mm) and then the duct is filled with a cementitious grout.

Investigations around the world have shown extensive problems with poor grouting of the ducts and consequent leakage of water, with and without chlorides, onto the steel cables with subsequent severe corrosion. In some cases this has led to failure of bridges and other structures (e.g. Woodward and Williams, 1988 and Hartt, 2019, 2020). The problem is that once the steel has been placed 50–100 mm or deeper within a structure, surrounded by a steel or polymer conduit, normal non-destructive test techniques are not effective.

Prestressing steel is loaded typically to around than 80% of its ultimate tensile strength. Therefore modest section loss, particularly due to pits, can lead to crack initiation and rapid, catastrophic failure. There are apocryphal tales of prestressing rods shooting out of buildings due to corrosion induced failures.

The following test techniques are used in these situations:

1   Radiography using mobile unit to 'X-ray' the structure and examine for tendon failures as discussed in Section 4.18.
2   Drilling in to ducts at two locations and using air flow to check for poor grouting.
3   A 'magnetic flux' technique that may be able to detect section loss and tendon failure using a high powered magnet and detecting magnetic flux leakage (Ghorbanpoor and Shi, 1995). This is still in development.
4   More conventional investigations of the anchorages and ducts to see if they are in good condition or whether any deterioration has occurred that could also have reached the tendons within the ducts.

The failures of prestressed concrete structures and concern about methods of investigating and rehabilitating them are a major issue in the civil engineering and concrete repair industries. Systems for monitoring post tensioned steel strands are discussed in Buenfeld *et al.* (2008). There is also a discussion of ways of designing post tensioning systems so that the ducts are isolated from the reinforcement and monitoring the electrical resistance between them to ensure the integrity of the ducts in FIB (2005). The problems of repairing and rehabilitating these structures are discussed in later chapters.

References

Alldred, J.C. (1993). Quantifying the Losses in Cover-meter Accuracy Due to Congestion of Reinforcement. *Proc. 5th Intl. Conf. on Structural Faults and Repair*, 2: 125–130.
Alongi, A.A., Clemena, G.G. and Cady, P. (1993). Condition Evaluation of Concrete Bridges Relative to Reinforcement Corrosion, Volume 3: Method for Evaluating the Condition of Asphalt Covered Decks. SHRP-S-325 Strategic Highway Research Program. National Research Council, Washington, DC.
American Concrete Institute (2008). Guide for Conducting a Visual Survey of Concrete in Service. Report by ACI Committee 201 ACI 201.1R-08. American Concrete Institute, Farmington Hills, MI.
American Concrete Institute (2019). Protection of Metals in Concrete Against Corrosion. Reported by ACI Committee 222. ACI 222R-19 American Concrete Institute, Farmington Hills, MI.
American Concrete Institute (2019). Design and Construction Practices to Mitigate Corrosion of Reinforcement in Concrete Structures. Reported by ACI Committee 222. ACI 222R-19. American Concrete Institute, Detroit, MI.
AMPP/NACE Report 01110 Stray-Current-Induced Corrosion in Prestressed and Conventionally Reinforced Concrete Structures. AMPP, Houston, TX.
AMPP/NACE SP21427-2019 Detection and Mitigation of Stray-Current Corrosion of Reinforced and Prestressed Concrete Structures. AMPP, Houston, TX.

Andrade, C., Alonso, M.C. and Gonzalez, J.A. (1990). An Initial Effort to Use Corrosion Rate Measurements for Estimating Rebar Durability. Corrosion Rates of Steel in Concrete, ASTM STP 1065, N.S. Berke, V. Chaker and D. Whiting (eds), ASTM, West Conshohocken, PA, pp 29–37.

Andrade, C., Alonso, C., Feliú, S. and González, J.A. (1995). Progress on Design and Residual Life Calculation with Regard to Rebar Corrosion on Reinforced Concrete. In Techniques to Assess the Corrosion Activity of Steel Reinforced Concrete Structures, ASTM STP 1276, N.S. Berke, E. Escalante, C.K. Nmai and D. Whiting (eds), American Society of Testing and Materials, West Conshohocken, PA.

Arup, H. and Klinghoffer, O. (1998). Junction Potentials at a Concrete/Electrolyte Interface. European Federation of Corrosion Publication. No. 25:31–39. ISSN: ISBN 1–86125–083–5.

ASTM C1202 (2019). Standard Test Method for Electrical Indication of Concrete's Ability to Resist Chloride Ion Penetration. American Society of Testing and Materials, West Conshohocken, PA.

ASTM C457-16 (2016). Standard Test Method for Microscopical Determination of Parameters of the Air-void System in Hardened Concrete. American Society of Testing and Materials, West Conshohocken, PA.

ASTM C856-20 (2020). Standard Practice for Petrographic Examination of Hardened Concrete. American Society of Testing and Materials, West Conshohocken, PA.

ASTM C876-15 (2015). Standard Test Method for Half-cell Potentials of Uncoated Reinforcing Steel in Concrete. American Society of Testing and Materials, West Conshohocken, PA.

ASTM D4788-13 (2013). Standard Test Method for Detecting Delaminations in Bridge Decks Using Infrared Thermography. American Society for Testing and Materials, West Conshohocken, PA.

ASTM D6087-15 (2015). Standard Test Method for Evaluating Asphalt-covered Concrete Bridge Decks Using Ground Penetrating Radar. American Society for Testing and Materials, West Conshohocken, PA.

ASTM G57-20 (2020). Field Measurement of Soil Resistivity Using the Wenner Four-electrode Method. American Society of Testing and Materials, West Conshohocken, PA.

ASTM G109 (2021). Standard Test Method for Determining the Effects of Chemical Admixtures on the Corrosion of Embedded Steel Reinforcement in Concrete Exposed to Chloride Environments. American Society of Testing and Materials, West Conshohocken, PA.

Beeby, A.W. (October 1985). Development of a Corrosion Cell for the Study of the Influences of Environment and Concrete Properties on Corrosion. Concrete 85, Conference, Brisbane, pp 118–123.

Bennett, J.E. and Mitchell, T.A. (1992). Reference Electrodes for Use with Reinforced Concrete Structures. Corrosion 92, Paper 191, Nashville.

Berke, N.S., Shen, D.F. and Sundberg, K.M. (1990). Comparison of the Linear Polarisation Resistance Technique to the Macrocell Corrosion Technique. Corrosion Rates of Steel in Concrete, ASTM STP 1065, N.S. Berke, V. Chaker and D. Whiting (eds), ASTM, Philadelphia, PA, pp 38–51.

BRE Digest 405 (1995). Carbonation of Concrete and its Effects on Durability. Building Research Establishment, IHS Markit Global, London.

BRE Digest 444 (2000). Corrosion of Steel in Concrete: Part 2 – Investigation and Assessment Building Research Establishment, IHS Markit Global, London.

BRE Information Paper IP 11/98 (1998). Assessing Carbonation Depth in Ageing High Alumina Cement Concrete. Building Research Establishment, Garston, UK, Publ CRC Ltd. London.

Broomfield, J., Davies, K., Hladky, K. and Noyce, P. (2003). Monitoring of Reinforcement Corrosion in Concrete Structures in the Field. Concrete Solutions, 1st Intern. Conference on Concrete Repair, St.-Malo, France, Proceedings. M. Grantham, F. Rendell, R. Jauberthie and C. Lanos (eds). Concrete Solutions (on CD Rom).

Broomfield, J.P. (2000). Results of Long Term Monitoring of Corrosion Inhibitors Applied to Corroding Reinforced Concrete Structures. Corrosion 2000. Paper No. 791.

Broomfield, J.P., Langford, P.E. and Ewins, A.J. (1990). The Use of a Potential Wheel to Survey Reinforced Concrete Structures. Corrosion rates of Steel in Concrete, ASTM STP 1065, in N.S. Berke, V. Chaker and D. Whiting (eds), ASTM, Philadelphia, PA, pp 38–51.

Broomfield, J.P., Rodriguez, J., Ortega, L.M. and Garcia, A.M. (1993). Corrosion Rate Measurement and Life Prediction for Reinforced Concrete Structures. *Proc. Structural Faults and Repair-93*, University of Edinburgh, Scotland, 2: 155–164.

Broomfield, J.P., Rodríguez, J., Ortega, L.M. and García, A.M. (1994). Corrosion Rate Measurements in Reinforced Concrete Structures by a Linear Polarisation Device. Philip D. Cady Symposium on Corrosion of Steel in Concrete, R.E. Weyers (ed.), American Concrete Institute Special Publication 151.

BS 1881 Testing Concrete. Part 124: 1988 Methods of Analysis of Hardened Concrete.

BS 1881–208 (1996). Testing Concrete – Part 208: Recommendations for the Determination of the Initial Surface Absorption of Concrete. British Standards Institute, London.

BS 7361 (1991). Cathodic Protection: Part 1. Code of Practice for Land and Marine Applications. British Standards Institute.

BS EN 1062–3 (2008). Paints and Varnishes – Coating Materials and Coating Systems for Exterior Masonry and Concrete. Part 3: Capillary Absorption and Permeability to water. British Standards Institute, London.

BS EN 14629 (2007). Products and Systems for the Protection and Determination of Chloride Content in Hardened Concrete. British Standards Institute.

BS EN 14630 (2006). Products and Systems for the Protection and Repair of Concrete Structures – Test Methods – Determination of Carbonation Depth in Hardened Concrete by the Phenolphthalein Method. British Standards Institute, London.

Buenfeld, N.R., Davies, R.D., Karimi, A. and Gilbertson, A.L. (2008). Intelligent Monitoring of Concrete Structures. CIRIA Report 661. CIRIA, London.

Bungey, J.H. (ed.) (1993). Non-Destructive Testing in Civil Engineering. International Conf. by The British Institute of Non-Destructive Testing, Liverpool University, UK.

Cady, P.D. and Gannon, E.J. (1992). Condition Evaluation of Concrete Bridges Relative to Reinforcement Corrosion. Volume 1: State of the Art of Existing Methods. National Research Council, Washington, DC. SHRP-S-330.

Chess, P. and Grønvold, F. (1996). Corrosion Investigation – a Guide to Half Cell Mapping. Thomas Telford, London.

Clear, K.C. (1989). Measuring the Rate of Corrosion of Steel in Field Concrete Structures. Transportation Research Board Preprint 324. 68th Annual Meeting, Washington, DC.

Concrete Bridge Development Group Technical Guide 2 (2002). Guide to Testing and Monitoring the Durability of Concrete Structures. The Concrete Society, Camberley, UK.

Concrete Society (1997a). Guide to Surface Treatments for Protection and Enhancement of Concrete. Technical Report 50. The Concrete Society, Camberley, UK.

Concrete Society (1997b). Guidance on Radar Testing of Concrete Structures. Technical Report 48. The Concrete Society, Camberley, UK.

Concrete Society in Association with the Federation of Resin Formulators and Applicators and the Association of Gunite Contractors (1984). Repair of Concrete Damaged by Reinforcement Corrosion. Report of a Working Party. The Concrete Society Technical Report No 26. The Concrete Society, Camberley, UK.

Concrete Society (2004). Electrochemical Tests for Reinforced Concrete. A Joint Concrete Society/Institute of Corrosion Report. Technical Report 60. The Concrete Society, Camberley, UK.

Dawson, J.L. (1983). Corrosion Monitoring of Steel in Concrete. Corrosion of Reinforcement in Concrete Construction, A.P. Crane (ed.). Ellis Horwood, Chichester, UK, pp 175–192.

Feliú, S., González, J.A., Andrade, C. and Feliú, V. (1988). 'On-Site Determination of the Polarization Resistance in a Reinforced Concrete Beam'. *Corrosion*, 44(10): 761–765.

FIB (2005). Durability of Post-Tensioning Tendons. Bulletin 33, FIB, Switzerland, Lausanne.

Fliz, J., Sehgal, D.L., Kho Y-T., Sabotl, S., Pickering, H., Osseo-Assare, K. and Cady, P.D. (1992). Condition Evaluation of Concrete Bridges Relative to Reinforcement Corrosion. Volume 2: Method for Measuring the Corrosion Rate of Reinforcing Steel. National Research Council, Washington, DC. SHRP-S-324.

Gannon, E.J. and Cady, P.D. (1992). Condition Evaluation of Concrete Bridges Relative to Reinforcement Corrosion – Volume 1: State of the Art of Existing Methods. SHRP/FR-92-103. National Research Council, Washington, DC.

Ghorbanpoor, A. and Shi, S. (1995). Assessment of Corrosion of Steel in Concrete Structures by Magnetic Based NDE Techniques. In Techniques to Assess the Corrosion Activity of Steel Reinforced Concrete Structures, ASTM STP 1276, N.S. Berke, E. Escalante, C. Nmai and D. Whiting (eds), American Society of Testing and Materials, Philadelphia, PA.

Glass, G.K. and Buenfeld, N.R. (2000). 'The Influence of Chloride Binding on the Chloride Induced Corrosion Risk in Reinforced Concrete'. *Corrosion Science*, 42: 329–344.

Gowers, K.R., Millard, S.G. and Gill, J.S. (1992). Techniques for Increasing the Accuracy of Linear Polarisation Measurement in Concrete Structures. Paper 205, NACE, Houston, TX.

Grantham, M. and Van Es, R. (1995). 'Admitting that Chlorides are Variable'. *Construction Repair*, 9(5): 7–10.

Grantham, M.G. (1993). 'An Automated Method for the Determination of Chloride in Hardened Concrete'. *Fifth Intl. Conf. on Structural Faults and Repair*, 2: 131–136.

Hart (2020). 'Corrosion Induced Failure of Bridge Post-Tensioning Tendons'. *Materials Performance* 57(10): 42–44.

Hartt, W.H. (2019). 'Failure Projection of Corroding Bridge Post Tensioning Tendons Considering Influential Factors and Issues'. *Corrosion* 75(9): 1146–1151.

Hausmann, D.A. (1967). 'Corrosion of Steel in Concrete: How Does It Occur?' *Materials Protection*, 6: 19–23.

Herald, S.E., Henry, M., Al-Qadi, I., Weyers, R.E., Feeney, M.A., Howlum, S.F., and Cady, P.D. (1992). Condition Evaluation of Concrete Bridges Relative to Reinforcement Corrosion. Volume 6: Method of Field Determination of Total Chloride Content. National Research Council, Washington, DC. SHRP-S-328.

Kaetzel, L., Clifton, J., Snyder, K. and Kleiger, P. (1994). Users Guide to the Highway Concrete (HWYCON) – Expert System. SHRP Report and Computer Program SHRP-C-406 National Research Council, Washington, DC.

Kropp, J. and Hilsdorf, H.K. (1995). Performance Criteria for Concrete Durability. Rilem Report 12, E&FN Spon, London, UK.

Langford, P. and Broomfield, J. (1987). 'Monitoring the Corrosion of Reinforcing Steel'. *Construction Repair*, 1(2): 32–36.

Lawler, J.S., Kurth, J.C., Garrett, S.M., Krauss, P.D. (2021). 'Statistical Distributions for Chloride Thresholds of Reinforcing Bars'. *ACI Materials J.* 118(2): 13–20.

Matthews, S.L. (1998). Application of Subsurface Radar as an Investigative Technique. CRC Ltd., London, BRE Report BR340.

Millard, S.G., Harrison, J.A. and Gowers, K.R. (1991). 'Practical Measurement of Concrete Resistivity'. *British Journal of NDT*, 33(2): 59–63.

NACE Standard Recommended Practice (2000). Cathodic Protection of Reinforcing Steel in Atmospherically Exposed Concrete Structures. RP0290–90 National Association of Corrosion Engineers, Houston, TX.

Neville, A.M. (1995). Properties of Concrete 4th Edition. Longman, London, pp 693–694.

Nustad, G.E. (Jul, 1994). Production and Use of Standardized Chloride Bearing Concrete Dusts for the Calibration of Equipment and Procedures for Chloride Analysis. Corrosion and Corrosion Protection of Steel in Concrete, Sheffield Academic Press, Sheffield, UK.

Polder (2000). 'Test Methods for the On Site Measurement of Resistivity of Concrete'. *Materials and Structures,* 33: 603–611.

Rapa, M. and Hartt, W.H. (Apr, 1999). Non-Destructive Evaluation of Jacketed Prestressed Concrete Piles For Corrosion Damage. NACE Corrosion 99. Paper No. 566.

Rodríguez, J., Ortega, L.M. and García, A.M. (1994). Asessment of Structural Elements with Corroded Reinforcement. Corrosion and Corrossion Potection of Steel in Concrete, R.N. Swamy (ed.), Sheffield Academic Press, Sheffield, UK, pp 171–185.

Shiessl, P. and Ruapach, M. (April, 1993). Non-Destructive Permanent Monitoring of the Corrosion Risk of Steel in Concrete. Non-Destructive Testing in Civil Engineering, J.H. Bungey (ed.), British Institute of Non-Destructive Testing, International Conference, University of Liverpool, 2: 661–654.

Stark, D. (1991). Handbook for the Identification of Alkali Silica Reactivity in Highway Structures. SHRP-C-315 Strategic Highway Research Program. National Research Council, Washington, DC.

Stern, M. and Geary, A.L. (1957). 'Electrochemical Polarisation. I. A Theoretical Analysis of the Shape of Polarisation Curves'. *J. Electrochem. Soc.*, 104: 56–63.

Titman, D.J. (1993). Fault Detection in Civil Engineering Structures Using Infra-Red Thermography, *Proc. 5th Intl. Conf. on Structural Faults and Repair*, 2: 137–140.

Tuutti, K. (1982). Corrosion of Steel in Concrete Swedish Cement & Concrete Research Institute, Stockholm.

Vassie, P.R. (1987). The Chloride Concentration and Resistivity of Eight Reinforced Concrete Bridge Decks after 50 Years Service. Research Report 93. Transport and Road Research Laboratory.

Whiting, D., Ost, B., Nagi, M. and Cady, P. (1992). Condition Evaluation of Concrete Bridges Relative to Reinforcement Corrosion, Volume 5: Methods for Evaluating the Effectiveness of Penetrating Sealers. SHRP-S-327, Strategic Highway Research Program. National Research Council, Washington, DC.

Woodward, R.J. and Williams, F.W. (1988). 'Collapse of Ynys-y-Gwas Bridge, West Glamorgan'. *Proc. Instn. Civ. Engrs. Part 1*, 84: 635–669.

Chapter 5

Corrosion monitoring

Chapter 4 dealt with the different test methods available for assessing corrosion related properties of reinforced concrete. Most of the techniques are applied with hand-held battery operated equipment and a 'one off' reading, or series of readings is taken. In some cases such as cover meter measurements that reading will not change. In others, such as concrete resistivity, corrosion rate or corrosion potential (reference electrode potential) it will change as aggressive agents in the environment move into the concrete and lead to or accelerate corrosion.

We can define corrosion monitoring as collecting corrosion related data on a regular basis. In this chapter we will exclude strain gauge monitoring which is adequately covered elsewhere in the literature, and cathodic protection monitoring which is discussed in Section 7.5.2.

Corrosion monitoring can be done on new or existing structures. Its application to new structures requires forethought on the part of the designer, a clear understanding of what is required, what sensors and monitoring equipment is available and where to site sensors. Probe installation is relatively straight forward and can be integrated with the construction process along with power, signal and monitoring systems. The term structural health monitoring has been invoked to describe the implementation of corrosion and other monitoring sensors on bridges and other structures, see Buenfeld *et al.* (2008), which includes specialist techniques for monitoring prestressed structures. A useful case history can be found in Webb et al. (2014).

In existing structures corrosion monitoring may be by regular inspection using the techniques described in Chapter 4, or by probes installed in the concrete or on its surface. These will have wired or wireless connections to monitoring equipment.

There are two major issues with corrosion monitoring:

1   What can be monitored?
2   What are the merits of monitoring?

DOI: 10.1201/9781003223016-5

We can monitor with a limited number of the techniques listed in Chapter 4. The main techniques are:

- concrete resistivity
- reference electrode potential
- corrosion rate.

In addition, there are relative humidity probes that can be embedded in concrete.

The reasons for carrying out monitoring are:

- early warning of significant deterioration (limit state being reached)
- input into planned maintenance
- monitoring the effectiveness of repairs or protective measures being installed
- development of a whole life deterioration and costing model for a structure or group of structures.

5.1   Regular surveys to monitor corrosion

This has been done on motorway bridges in the United Kingdom where corrosion caused by deicing salt leakage through expansion joints onto substructures is an acknowledged problem. The major cost is access on each survey. The progress in chloride ingress and reference electrode potentials was monitored along with a delamination survey.

Broomfield (2000) carried out regular surveys using LPR (see Section 4.12) on some reinforced concrete support pillars which had calcium chloride cast in as a set accelerator. These had been repaired with vapour phase corrosion inhibitors applied. Measurements from 1995 to 1999 showed the increase in the corrosion rate in treated and untreated control areas, with the corrosion rate peaking at around 0.5 to 1.0 $\mu A/cm^2$ (5–10 mm per year) with cracking seen approximately one year later. This demonstrated that the inhibitor applied with the chloride content of the structures was too high to stop corrosion. It also showed that a factor of two difference in corrosion rate between treated and untreated area made no significant difference to the rate and extent of cracking.

5.2   Permanent corrosion monitoring systems

A number of the techniques described in Chapter 4 are suitable for embedding in concrete for permanent monitoring. The main requirements are that they are durable when cast in concrete and exposed to the environment, the readings are meaningful and that readings do not drift with time as recalibration is difficult or impossible.

5.2.1   Permanent reference electrodes

The simplest probe is the reference electrode. These are designed for exposure in concrete for installation with cathodic protection systems (Section 7.3). However, it has been shown that once embedded in concrete they cannot be recalibrated if they drift (Ansuini and Dimond, 1994) and a very large number are required if a useful 'potential map' (Section 4.8.4) is to be produced. Reference electrodes are incorporated into LPR probes (Section 5.2.2) but are rarely used on their own.

As stated in Section 4.12, corrosion rates can be measured by linear polarization or by galvanic or macrocell techniques. Both have their merits and limitations as discussed in Sections 5.2.2 and 5.2.3.

5.2.2   Corrosion rate by polarization resistance sensors

The LPR probe is ideally suited to permanent installation both in new construction and as a retrofit into existing structures. Probes consist of a reference electrode, an auxiliary electrode and a working electrode. In the case of a 'new build' probe the working electrode can be a piece of steel of known surface area so that an accurate corrosion rate measurement is made on an electrode in the same environment as the rest of the reinforcement. An electrical connection to the reinforcement means that measurements can also be taken on the reinforcement itself. Figure 5.1 is an example of such a probe installed in a reinforcing cage prior to casting. The results from the installation are shown in Figure 5.3.

Such a design cannot be used in existing structures. In such cases the working electrode must be the reinforcement in an undisturbed area of concrete. This means that the probe consists of a reference electrode and

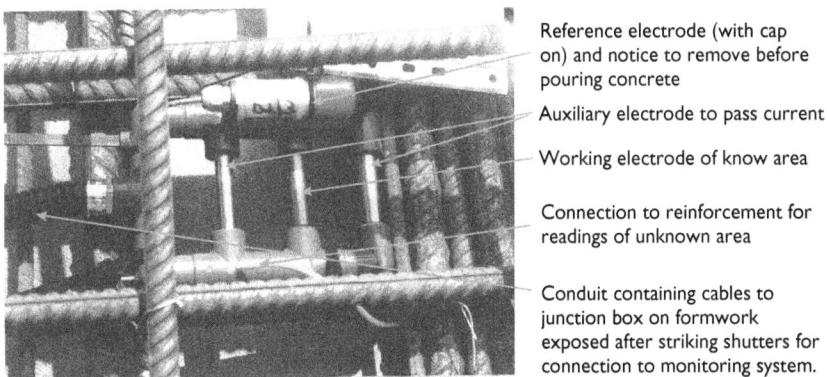

Reference electrode (with cap on) and notice to remove before pouring concrete

Auxiliary electrode to pass current

Working electrode of know area

Connection to reinforcement for readings of unknown area

Conduit containing cables to junction box on formwork exposed after striking shutters for connection to monitoring system.

Figure 5.1 A permanent LPR monitoring probe being installed in a precast deck unit of the Dartford Tunnel (Broomfield *et al.*, 2003).

Figure 5.2 A corrosion monitoring probe consisting of a reference electrode and an auxiliary electrode potted up in mortar in the process of installation into a continuously reinforced concrete pavement where corrosion monitoring is required due to the high level of chloride found in the mix water after laying several kilometres of concrete (Broomfield *et al.*, 2003) photograph.

auxiliary electrode with a connection to the reinforcement at a suitable location. The probe assembly must be installed with minimum disturbance of the reinforcement as shown in Figure 5.2.

5.2.3   Corrosion rate by galvanic sensors

A galvanic ladder probe for installation in new structures is shown in Figure 5.4. This consists of a series of mild steel anode rungs connected to a stainless steel cathode via an ammeter. When chlorides or carbonation reach successive rungs the current between anode and rises.

The ladder probe is designed for installation in new structures and has been installed in major bridges, tunnels and other structures throughout Europe. Variations on this design are supplied by different manufacturers.

A more recent development is a 'washer probe' that can be fitted tightly into a cored hole to provide similar data on existing structures as shown in Figure 5.5.

5.2.4   Corrosion rate by electrical resistance sensors

In an electrical resistance (ER) probe, a strip, wire or tube sensor of known cross section is exposed to the environment. In the case of concrete structures, it must be embedded at the time of construction otherwise it is not in the same environment as the reinforcement it is supposed to simulate. The sensor metal must be similar to the structure metal. The electrical resistance of the sensor is measured after initial installation and at subsequent time periods. As metal is lost from the exposed surface of the sensor, the measured electrical

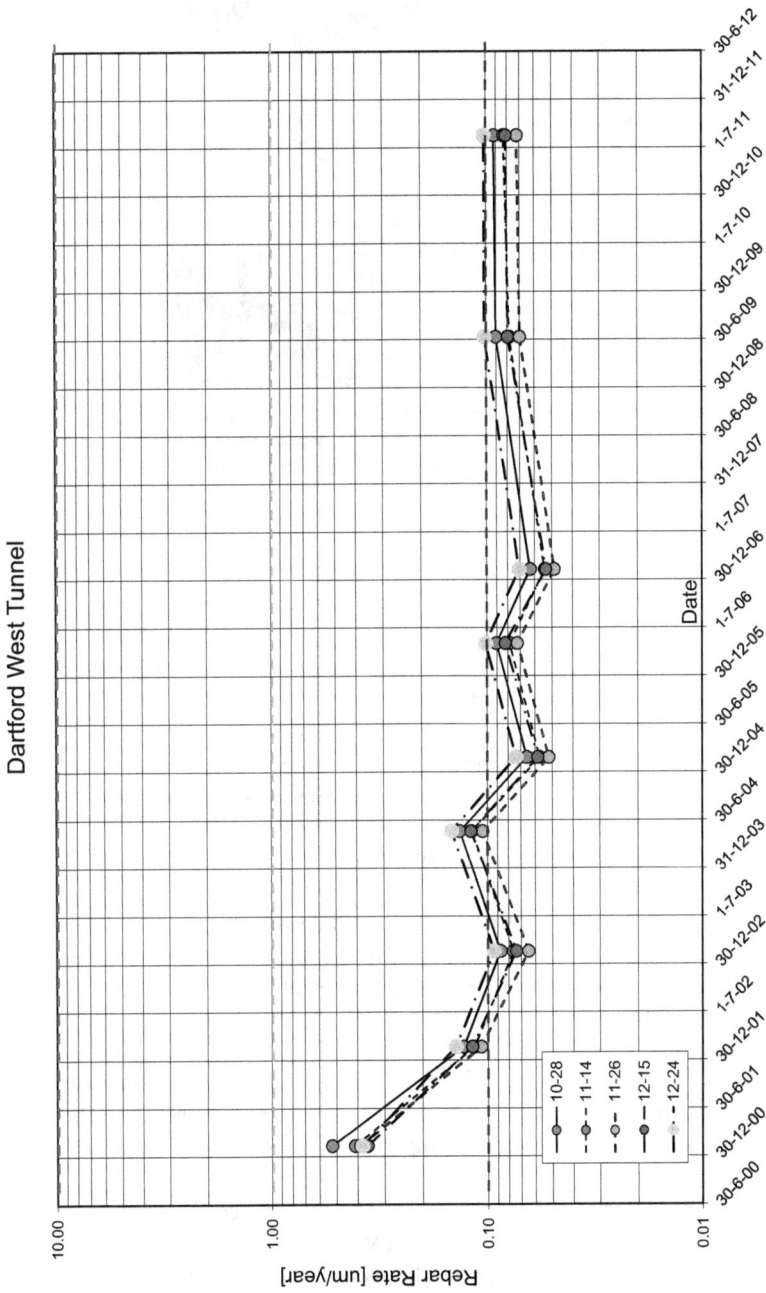

Figure 5.3 Approximately 12 years of LPR corrosion rate data from the probes shown in Figure 5.1 described in Broomfield et al., 2003.

Figure 5.4 Galvanic ladder probe (Raupach and Schiessl, 1995). Ladders are at an incline through the cover so that successive rungs depassivated and a galvanic current starts to flow between the anode and a noble metal cathode. Courtesy Dr-Ing. M. Raupach.

resistance will increase which allows the amount of loss to be quantified. A proprietary automated corrosion probe reader is used to report the thickness loss directly. This uses a Wheatstone bridge with dummy leads to compensate for the cables and uses an embedded thermometer to compensate for the effect of temperature on readings.

This is the only technique that can measure in situ corrosion loss under cathodic protection. Valid measurements are possible even in non-conductive environments. This technique measures losses from corrosion due to ineffective cathodic protection.

The main disadvantage of ER probes is that they only give valid data when the corrosion mode is uniform. These instruments are not suitable when corrosion is localized (pitting, cracking). This is its major limitation in concrete where corrosion initiates with pitting, particularly when chlorides are present. There is a trade-off between the sensitivity of the probe and its usable lifetime. The thinner the probe the more sensitive but the more quickly consumed.

Probes are permanently installed and measurements are made on a periodic basis. As a guide, monitor the probe at least once a month for the first 12 months. Thereafter, the monitoring frequency can be reduced to once every three months.

A graph is made of metal loss vs. time. The interval corrosion rate would be the slope of the graph between any two data points while the average corrosion rate would be the slope of the trendline as calculated by the least squares method. Many graphing programs, such as MS Excel, have built-in capability to determine trendline slopes.

Figure 5.5 (a) and (b) Galvanic 'multi ring electrode' (Raupach and Schiessl, 1995). The assembly shown in (a) can be fitted into a cored hole in the concrete cover or (b) can be cast into new concrete so that successive washers depassivated as the washers successively depassivate and a galvanic current starts to flow between the anode and a noble metal cathode. Courtesy Dr-Ing. M. Raupach.

When used to demonstrate the effectiveness of a cathodic protection system, an average corrosion rate (trendline) of less than 0.1 mils/year (2.5 microns/year) over a 12-month monitoring period indicates that cathodic protection is effective at the location of the sensor.

Corrosion probes of the electrical resistance type are rarely used in concrete. They can suffer from localized pitting around the ends of the corrodible steel leading to rapid failure once corrosion initiates in this environment. They have been used to show that corrosion is under control, for example, in impressed current cathodic protection systems, but in other cases have been found to be poor indicators of rates of corrosion.

5.2.5 Concrete resistivity sensors

Resistivity probes can be embedded in concrete. The design shown in Figure 5.6 was installed in the Dartford Tunnels on the River Thames estuary (Broomfield *et al.*, 2003). Data collected is shown in Figure 5.7.

5.2.6 Humidity monitoring

Commercially available relative humidity probes are available that are durable and are suitable for embedding in concrete. Section 4.12.4 discusses the effect of relative humidity on corrosion rate.

5.2.7 Chloride content and pH monitoring

There is discussion in the literature about monitoring pH and chloride content (McCarter *et al.*, 2004). However, there is no field data from such sensors and interpretation is difficult as the corrosion threshold is not a fixed value of pH or chloride content as discussed in Section 3.2.3.

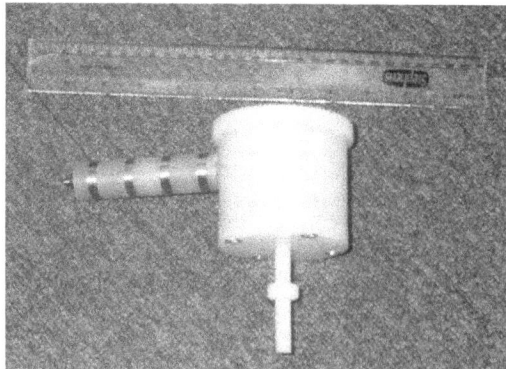

Figure 5.6 This resistivity probe is directly attached to the formwork so that when the concrete is cast around it the four stainless steel washers can be used to make four probe resistivity measurements.

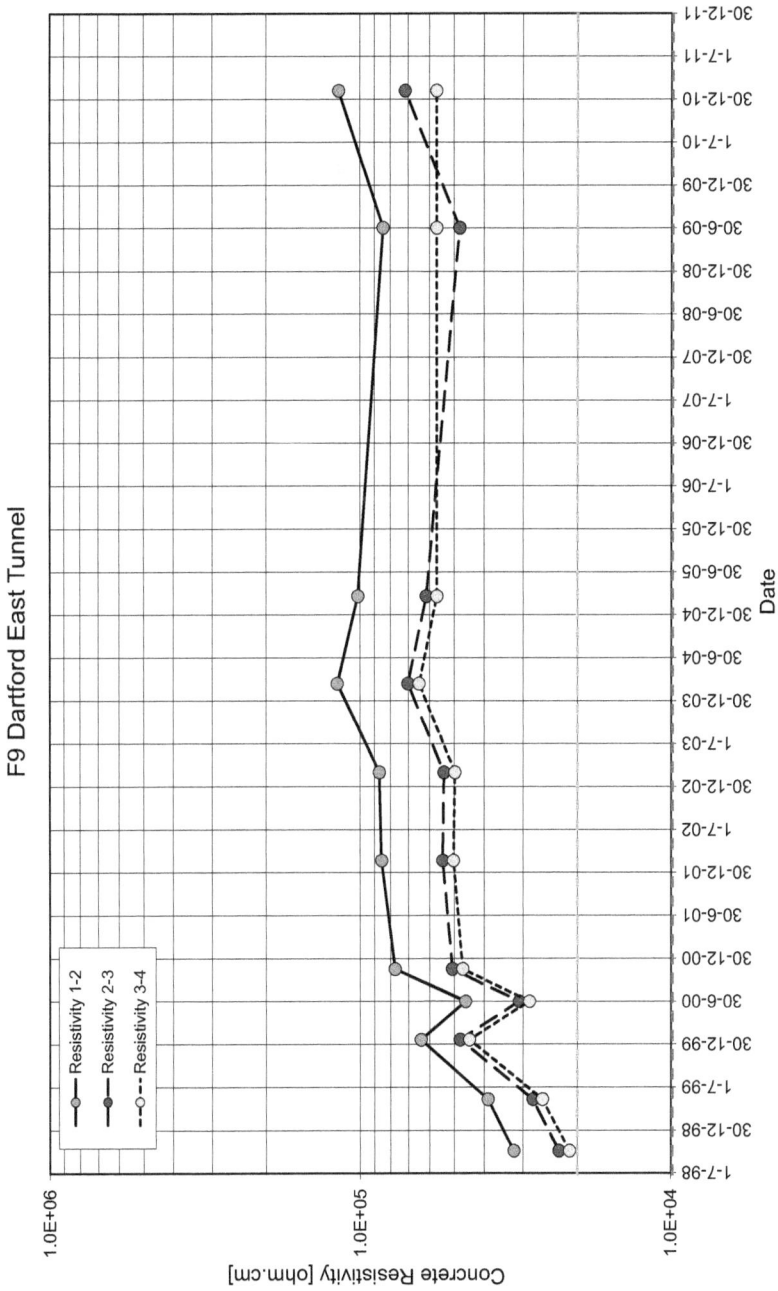

Figure 5.7 Approximately 12 years of resistivity data from probes in Figure 5.1 from deck units in the Dartford Tunnel, England (Broomfield et al., 2003).

Figure 5.8 Termination box with connections for reference electrodes, LPR probes and resistivity probes on a jetty.

5.3   Remote monitoring systems and data management

There is a range of options for a corrosion monitoring system: manual systems, local logging systems, remote monitoring and wireless. For easily accessible systems with a limited number of probes, a simple termination box with sockets for plugging in metres is suitable. Figure 5.8 shows a manual corrosion monitoring on a jetty with a LPR corrosion rate meter, a resistivity meter and a digital voltmeter for reading the reference electrode potentials (Broomfield, 1998). The data in Figure 5.7 was collected manually.

Data can be collected automatically and stored either for download to a portable computer brought to site or by telecoms connection where the engineer can connect remotely to the computer and download data (Lynch, J.P., Loh, J., 2006). The data in Figure 5.3 was collected on a central computer from networked sensors and downloaded remotely (Broomfield *et al.*, 2003). A range of systems for stuctural health monitoring, including corrosion monitoring, is discussed in Buenfeld *et al.* (2008). See also NACE (2007).

References

Ansuini, F.J. and Dimond, J.R. (1994). Long Term Stability Testing of Reference Electrodes for Reinforced Concrete. Corrosion 94. Paper No. 295.

Broomfield, J.P. (1998). 'Corrosion Monitoring'. *Concrete Engineering International*, 2(2): 27–30.

Broomfield, J.P. (2000). Results of Long Term Monitoring of Corrosion Inhibitors Applied to Corroding Reinforced Concrete Structures. Corrosion 2000. Mar; Paper No. 791. NACE, Houston, TX.

Broomfield, J., Davies, K., Hladky, K. and Noyce, P. (2003). Monitoring of Reinforcement Corrosion in Concrete Structures in the Field. Proceedings of Concrete Solutions, 1st International Conference on Concrete Repair, St. Malo, France. M. Grantham, F. Rendell, R. Jauberthie and C. Lanos (eds.), Concrete Solutions (on CD ROM).

Buenfeld, N.R., Davies, R.D., Karimi, A., Gilbertson, A.L. (2008). Intelligent Monitoring of Concrete Structures. CIRIA Report C661. CIRIA, London.

Lynch, J.P., Loh, J. (2006). 'A Summary Review of Wireless Sensors and Sensor Networks for Structural Health Monitoring'. *The Shock and Vibration Digest*, 38(91): 91–128.

McCarter, W.J. and Vennesland, O. (2004). 'Sensor Systems for Use in Reinforced Concrete Structures'. *Construction and Building Materials*, 18: 351–358.

NACE (2007). Report on Corrosion Probes in Soil or Concrete. Publication 05107. AMPP, Houston, TX

Raupach, M. and Schiessl, P. (Feb, 1995). Monitoring System for the Penetration of Chlorides, Carbonation and the Corrosion Risk for the Reinforcement. Proceedings of the 6th International Conference on Structural Faults and Repair.

Webb, G., Vardanega, P.J., Fidler, P. and Middleton, C. (2014). 'Analysis of Structural Health Monitoring Data from Hammersmith Flyover'. *Journal of Bridge Engineering*, 19:6.

Chapter 6

Physical and chemical repair and rehabilitation techniques

This chapter discusses the options for rehabilitating our corrosion damaged reinforced concrete structures. We will start by examining the conventional repairs by concrete removal and replacement. Removal techniques such as percussion tools, hydrojetting and milling and their limitations are discussed. The problems of conventional repair such as continued corrosion and structural considerations are covered, along with coatings, sealers, membranes and barriers. Encasement, overlays, corrosion inhibitors are also covered in terms of corrosion performance. Mix designs and concrete chemistry are not be discussed in detail as that information is better covered elsewhere (see, for instance, SHRP-C-345 *Synthesis of current and projected concrete highway technology* (Whiting *et al.*, 1993) or Emberson and Mays, 1990.

Electrochemical techniques such as cathodic protection, chloride removal and re-alkalization are discussed in a separate chapter following this one. Rehabilitation methodology is discussed in Chapter 8 after discussion of the major rehabilitation techniques.

The appropriate repair and rehabilitation system must be chosen for each structure according to its type, condition and its future use. There is no point in spending large amounts of money on a structure due for demolition or other major works in a few years. Equally it is pointless carrying out cosmetic patching to spalled concrete on a structure expected to last another 20–50 years where future access is difficult to arrange. What looks acceptable on a bridge is not necessarily acceptable on a building, where aesthetics have an important role to play, requiring good finishes to repairs and cosmetic coatings. The selection of suitable repairs is discussed in Chapter 8. The new European standard on concrete repair materials EN1504 will be mentioned in this chapter where relevant but is covered in more detail in Chapter 8.

One definition of the word rehabilitate is 'to restore to proper condition'. To repair is defined as 'to replace or refix parts, compensating for loss or exhaustion'. These definitions are worth bearing in mind. If we want to rehabilitate a structure we want to restore it, not necessarily to its original

DOI: 10.1201/9781003223016-6

condition, because if we do, it may fail again because of intrinsic flaws. We want to establish its 'proper' condition, that is, resistant to corrosion. In other words, to rehabilitate the structure we may need to improve it compared to its original condition. To repair is merely fixing the damage. This implies that deterioration may continue. Patch repairs are just what they say. They repair the damaged concrete. They will not stop future deterioration and may, as we will see later, accelerate it. Cathodic protection and the other electrochemical techniques can rehabilitate the structure. They mitigate the corrosion process across the whole treated area. Coatings and barriers can also rehabilitate if applied appropriately and at the correct time.

It is important to repeat that we are only discussing reinforcement corrosion repairs and rehabilitation. Other problems, structural defects, alkali-silica reaction (ASR) and so on may also need to be addressed at the same time and rehabilitation systems must be an integrated package of compatible, complementary systems. Some of the issues of compatibility are discussed later. Specifically the compatibility of patch repair systems and ASR with electrochemical treatments will be addressed.

Whatever repair or rehabilitation option we choose, most investigations start after concrete has already been damaged by corrosion. Some concrete removal and repair is therefore required on most jobs regardless of the rehabilitation or corrosion control technique selected.

6.1 Concrete removal and surface preparation

There are a number of methods for removing concrete and the choice depends on the specification, budget and contractor's preferences. If concrete is just starting to spall due to carbonation or if an electrochemical treatment like cathodic protection is planned then a simple repair of removing unsound concrete, cleaning the rebar surface, squaring the edges and putting in a sound, cementitious, non-shrinking repair material may suffice.

Before patching carbonated concrete, the cracked and spalled concrete must be removed from around the rebar or as deep as the carbonation front goes. The cementitious patch material is chosen to ensure that the steel is back in a high pH, alkaline environment. This will encourage the reformation of the passive layer to stop further corrosion.

If patching is required due to chloride-induced corrosion then the usual specification is to remove concrete to about 25 mm behind the rebar ensuring that all corroded steel is exposed around delaminated areas and the rebar is cleaned to a near white finish to remove all rust and chlorides. Figure 6.1 shows three types of repair: A bad repair, a good chloride repair and a patch for cathodic protection. The cut edges and faces of the concrete must be square and clean of all dust and debris. There is currently some discussion of whether the best bond between the original and patch repair material is achieved by slightly angling the repair edge and roughening it rather than a square cut so that the mortar fully fills the break out and bonds well.

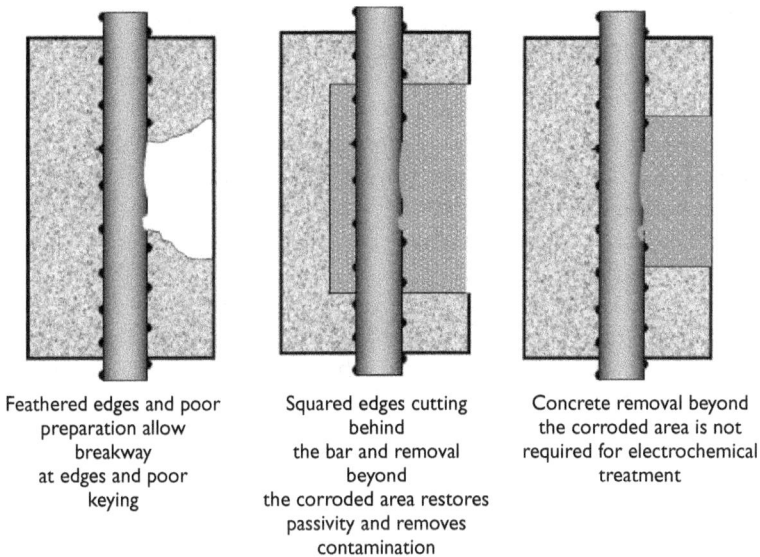

| Feathered edges and poor preparation allow breakway at edges and poor keying | Squared edges cutting behind the bar and removal beyond the corroded area restores passivity and removes contamination | Concrete removal beyond the corroded area is not required for electrochemical treatment |

Figure 6.1 Patch repairs, bad, good and compatible with electrochemical treatment.

The choice of patch repair material is discussed later. Wetting or bonding agents may be used to promote adhesion between the repair and the original concrete. Bonding agents should be checked for compatibility when being used where electrochemical treatment is required as it could create an insulating layer. If an electrochemical technique is being used then repairs only need to reinstate damaged concrete. A clean rebar surface is needed and a simple patch to rebar depth in cracked, spalled and delaminated areas will prepare the structure for the application of the anode. In some cases the anode and overlay will fill the excavated areas as a single operation. Materials used must be compatible with the electrochemical treatment. These issues are more fully discussed in Chapter 7.

In North America some highway agencies selectively remove the cover concrete in all areas where the half cell potentials exceed a threshold such as −250 mV against copper/copper sulphate, CSE). It has been shown (Weyers *et al.*, 1993) that the higher (less negative) the threshold the longer lasting the repair but this must be balanced by the cost and structural implications of extensive concrete removal.

6.1.1 *Percussion tools*

A summary of the properties and categories of hand held pneumatic breakers is given in Vorster *et al.* (1992). These have in many cases been superseded by electric tools by major manufacturers such as Hilti. For concrete repair work breakers usually range from about 10 to 45 kg, or a maximum of about 20 kg for vertical or overhead work. Small self contained electric units can be used for small areas in low strength concrete but pneumatic units with a separate compressor are required for large volume work.

Pneumatic hammers are labour intensive but have low capital cost and are very versatile. They will cut behind the bars and get between them. Research by the Strategic Highway Research Program (SHRP) has shown that the contract terms for a repair job are a major influence upon the contractor's decision as to how concrete removal is done (Vorster, 1992). If concrete removal is in small areas, pneumatic hammers are more practical than the larger plant such as hydrojetting or milling equipment discussed later. If the area of concrete to be broken out is undefined at the tender stage and the contractor is paid by the square metre (or cubic metre) there is less risk with a low capital cost approach like pneumatic hammers.

Breakers will not clean the rebars and are operator sensitive in the finish achieved. Work at the UK Transport Research Laboratory (Vassie, 1987) has shown that inadequate rebar cleaning will allow corrosion to proceed. Hydrojetting or grit blasting the rebars may be needed to remove all chloride contamination as discussed below.

Typical production rates for pneumatic breakers range from about 0.025 to 0.25 m^3 per hour for breakers ranging from 10 to 45 kg. Breakers are well suited for jobs with small, discontinuous areas of concrete removal, but not for large scale removal.

The requirements in health and safety and construction design in the United Kingdom and other countries require the minimization of the use of tools with high vibration due to the risk of hand arm vibration syndrome or 'vibration white finger' to operatives. Most specialist concrete repair contractors, at least in the UK, will have protocols to protect operatives when using percussion tools.

6.1.2 Hydrojetting

This is an increasingly popular method of removing concrete on decks and on substructures. In theory, on a deck, a vehicle mounted system can be set up to remove concrete to a uniform depth across large areas. It will run across the deck removing unsound concrete, cleaning the rebars and leaving a surface ready for patching and overlaying. There is minimal damage to sound concrete which means that this technique can be very cost effective on medium to large areas. In practice concrete quality often varies and therefore the amount and depth of concrete removed will also vary. Follow-up with pneumatic hammers can be required where the insufficient concrete is removed.

A small 'peak' or 'shadow' often remains behind the rebar after hydrojetting as shown schematically in Figure 6.2. If the specification or

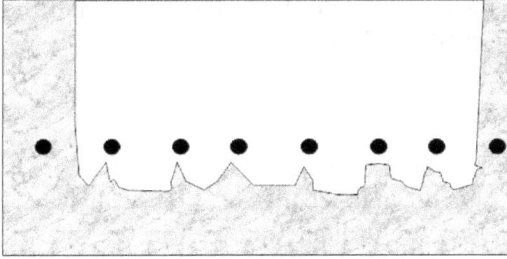

Figure 6.2 Schematic of results of hydrojetting showing the 'shadow' or peaks behind the reinforcing bars.

the engineer on site requires its removal then the economic and efficiency gains of using hydrojetting can be lost. Either the client or the contractor will end up with doubled costs for concrete removal. Therefore, if hydrojetting is required or preferred in a contract the contract specifications must reflect the performance of the technique. The required finish and the amount of concrete removed must reflect the performance of the hydrojetting technique, not a theoretical standard for preparation prior to repair.

Hydrojetting of substructures can be done manually by an operator holding a high pressure hose, or by a unit mounted on a robotic arm. Manual jetting requires a very high level of protection for the operator and for bystanders.

In all cases attention must be paid to the water run off. There is a risk of contaminating groundwater and streams with alkali, fine debris and the contaminants that have built up on the concrete surface. It may be necessary to put bunds around the repair area and collect the contaminated water or to filter the runoff, collect the solid particles and prevent them from being washed away. The water itself can undermine soil embankments or other structures unless runoff is controlled. The development of low water volume hydrojetting systems has reduced the problems of controlling runoff.

Hydrojetting units typically consist of an engine driving a high pressure pump connected via a high pressure hose to the nozzle which is either manually held or moves under microprocessor control within a protective guard system on a vehicle mounted system. Water is delivered at pressures ranging from about 80 to 140 MPa at flow rates ranging from 75 to 270 l/min. Hydrodemolition contractors protect information about production rates of their units. Typical rates will range from 0.25 m³/hr for small single pumped systems to 1 m³/hr for top of the range dual pumped systems. This assumes a 28 MPa concrete and removal down to 75 mm. American hydrojetting practices are discussed in Vorster *et al.* (1992). European practices are reviewed in Ingvarsson and Erikkson (1988) and Anon (1993). The Swedish Bridge Code 88 section 7 on maintenance, repair and strengthening discusses hydrodemolition and there is an ICRI guide on the subject (ICRI, 2004).

6.1.3 Milling machines

Milling machines can be used to remove concrete cover on decks. They must not be used right down to rebar level or they will damage the bars and the milling head. They are of limited interest in the United Kingdom and Europe with respect to corrosion repairs on bridge decks due to the extensive use of waterproofing membranes, but extensively used to mill off the asphalt wearing course with or without the waterproofing layer. They are widely used in North America where heavily contaminated concrete bridge decks are frequently milled with local removal to expose corroding rebars that have delaminated the concrete.

Milling machines are far more precise in removing a defined amount of cover compared with hydrojetting. However, they will cause considerable damage to themselves and to the deck if they catch tie wires or stirrups, so great care must be taken to ensure that the cover is uniform and a margin is allowed to prevent exposure of the steel work.

Removal rates can be very rapid, usually greater than 1 m³/min using machines with a 2 m wide cutting mandrel travelling at 1.5 m/min.

After milling, local, deeper removal is needed where cracking and spalling has occurred. The deck is then patched in the delaminated areas and a dense cementitious overlay is put back on. The use of concrete overlays is discussed in a later section in this chapter.

6.1.4 Comparative costing

Little work has been carried out on comparative costing. Partly this is because of the difficulty of getting comparative values, the commercial sensitivity of the rates used and the difficulty of making comparisons that are valid for more than one job.

Figure 6.3 shows a graph comparing milling, hydrojetting and jack hammering based on 1860 m² of concrete removal to a depth of 64 mm on a bridge deck in the United States (Vorster et al., 1992). Using this analysis, breakers are never economic even for an area of 50 m². This suggests that capital costs of equipment have not been compared adequately for small areas. However, it does serve to illustrate the commercial benefits of using the high tech approach.

6.1.5 Concrete damage and surface preparation

There has been some discussion in the technical literature (Vorster et al., 1992) about damage done to the parent concrete during concrete removal

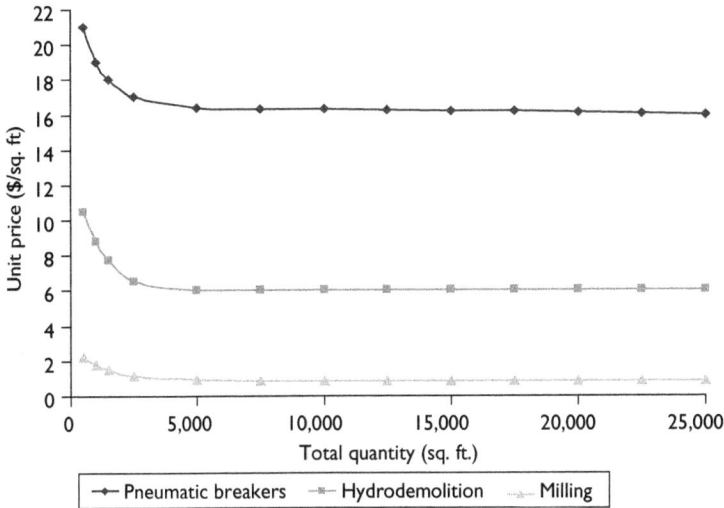

Figure 6.3 Comparison of costs for concrete removal from Voster *et al.* (1992). Courtesy Transportation Research Board (Strategic Highway Research Program).

especially by milling and breakers. There is no clear indication of the effect of such damage on repairs. However, a damaged substrate is more likely to deteriorate more rapidly and be more permeable to moisture and chloride ingress.

For the best bonding between old and new concrete a rough, clean and crack-free surface is required. The surface should be damp enough not to suck moisture out of the new concrete which is needed for hydration. However, the surface must not be saturated or ponded with water as this will increase the water/cement ratio of the new concrete and weaken it and the bond between the parent and the repair material.

The exception to this is when bonding agents are used. These must be applied according to the manufacturers' instructions, particularly where there is a requirement to place the new concrete while the bonding agent is still 'tacky'. This is often very difficult to ensure under site conditions and requires good supervision and quality assurance on site.

Hydrojetting is probably the concrete removal technique of choice for minimal damage, cleaning of rebars and removal to required depth. In all cases the edges of removed areas must be square to the surface with no 'feathering' that would be difficult to fill with reasonably sized aggregates (Figure 6.2). However, there is recent consensus by concrete repair specialist contractors that sloping edges with a wider opening minimise the risk of poor consolidation and bonding of the repair mix to the parent concrete.

6.2 Patches

Having removed the damaged and contaminated concrete we must patch it. Many proprietary patch materials are on the market. Such pre-bagged materials are most likely to be applied properly, especially to small repair areas, but they are more expensive. If repair contractors must measure quantities and mix on site it will save money at the risk of less consistency and higher risks of shrinkage, poor bonding, etc. Specialized mix design can be carried out by concrete experts to provide pumpable, pourable and trowelable mixes. In the United States and Canada many state highway agencies have developed their own mix designs for concrete repair work based on locally available materials and the prices of additives and cement replacement materials such as microsilica, polymers, water reducing agents, etc.

CIRIA Technical Note 141 (CIRIA, 1993) laid down basic requirements of shrinkage, expansion due to wetting and temperature, modulus of elasticity, creep, etc. These have been taken up in standards such as the European Standard on concrete repair BS EN 1504 Products and repair systems for the protection and repair of concrete structures – definitions, requirements, quality control and evaluation of conformity Parts 1 to 10. This is discussed more fully at the end of this chapter.

Most proprietary, pre-bagged mixes carry guarantees of the materials particularly for carbonation repairs. Manufacturers and applicators will be more cautious with chloride repairs and cannot guarantee that all chloride is removed in areas adjacent to the patch. The incipient anode problem (see Section 6.2.1) is far more prevalent in chloride contaminated structures so corrosion will continue around the patches.

Fewer and fewer materials manufacturers have their own company applicators although most have approved or recommended applicators. It is now possible to get 'back to back' guarantees for concrete repairs to last 5–15 years, although there is a price to pay for the premium on such insurance backed guarantees.

Some proprietary patch repair materials include bonding agents. These must be checked for compatibility with electrochemical rehabilitation work. They may help where a high standard of workmanship is difficult to achieve, but they often require that the patch is applied at the right time when the bonding agent is ready. This can be difficult to do on site and adequate supervision is necessary to ensure that correct application is achieved.

Another major issue is curing. This can be done with curing membranes, wet hessian, plastic sheeting. Well cured repairs will perform far better than those left to 'air cure' especially in hot or windy weather when there is a risk of rapid drying.

6.2.1 Incipient anodes

We have already seen that patch repairing is not usually adequate to stop further deterioration in the presence of chloride attack. If a structure with

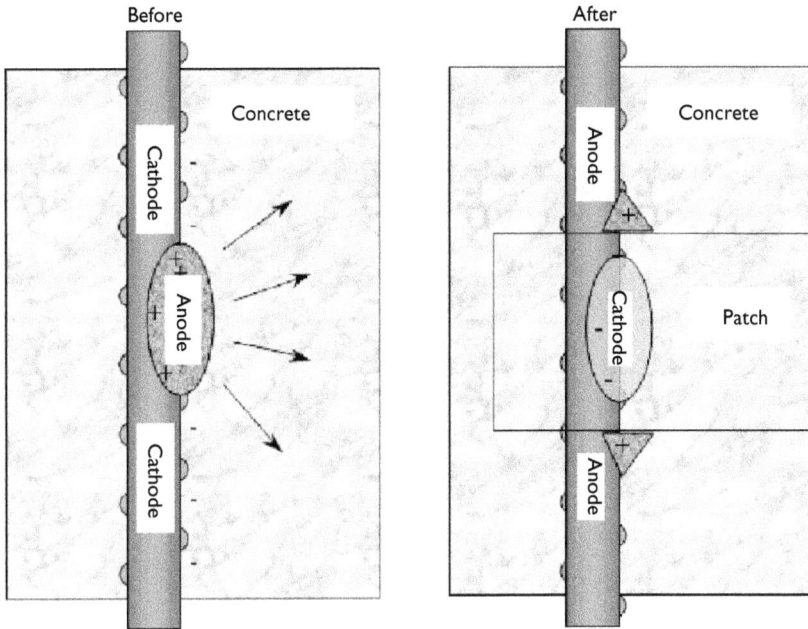

Figure 6.4 Incipient anode schematic showing how anode is displaced to the edge of the repair by the formation of a new cathode in the patch repair.

extensive chloride attack is to be patch repaired then it must be recognized that patching the corroding areas can accelerate corrosion elsewhere. When we stop the anodic reaction (2.1) we stop the generation of hydroxyl ions at the cathode (equation (2.2)). Therefore, areas previously protected from corrosion because they were made cathodic by the proximity of the anode (now repaired) will rise above the critical chloride/hydroxyl ratio and corrosion will be initiated. This often occurs around the new patch as shown schematically in Figure 6.4. This 'incipient anode' problem is avoided by applying an electrochemical rehabilitation technique.

One of the first cathodic protection installations in the United Kingdom was to a police station in the North of England. The author visited the station in 1987 and was shown the state of corrosion and the specification of a high quality, high cost patch repair and epoxy coating applied across the portal arches. A year later the author was invited back to examine the corrosion around the patches and it was agreed that the patches and the coating would be removed and a cathodic protection system would be installed. All the previous work had to be undone as the repairs would be incompatible with impressed current cathodic protection. The problem was incipient anodes forming around the patches. This story of incipient anodes forming around high quality patch repairs has been repeated on many structures particularly in marine and deicing salt environments. Figure 6.5 shows two examples of incipient anodes around repairs.

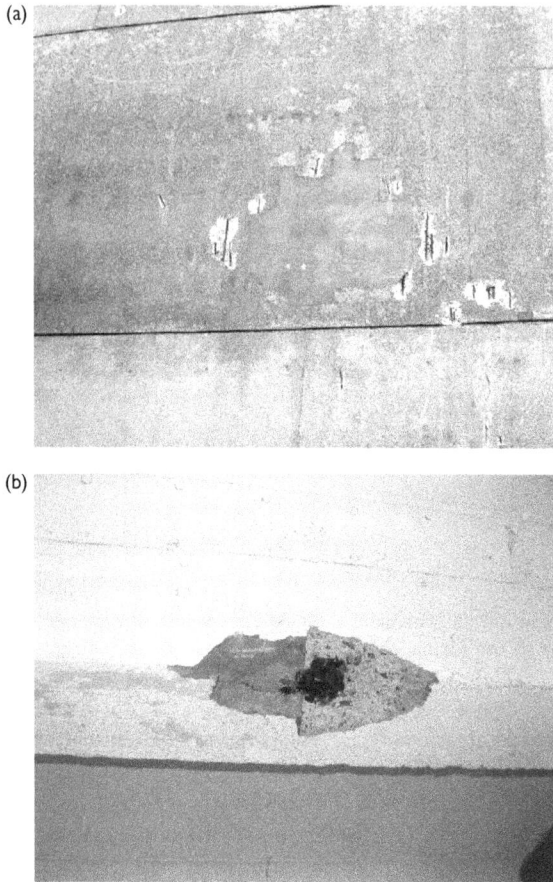

Figure 6.5 (a) Patch repair with surrounding spalling due to incipient anodes on a building. (b) Incipient anode showing repair (left side) and corrosion in original concrete.

6.2.2   Load transfer and structural issues

Removing the concrete cover, either by spalling or concrete removal, redistributes the load within the structure. The exposed steel may bend once the bond between steel and concrete is lost. This can happen due to corrosion as well as during repair. Obviously, bending of the steel will give severe structural problems. Any significant concrete removal or corrosion damage must be assessed by a structural engineer and propping may be necessary during repairs, particularly on substructures or any other load bearing element.

It cannot be assumed that a patch repair will take load in the same way as the original concrete. This can be particularly important for slender, lightly reinforced elements on buildings. Cutting out concrete behind the rebar can be a risky operation and a patched structure, although cosmetically attractive (particularly if a coating is applied), may have lost considerable load bearing capacity.

The author has been involved with repairs to several structures where there was a risk of loss of concrete to reinforcement bond potentially leading to bars bending when exposed for patching due to chlorides. This led to either an inferior patch repair (because chloride laden concrete was not removed from behind the rebar) or a decision to cathodically protect the structure to avoid the need to carry out such strenuous repairs (see Figure 6.1). An example of a load bearing system being used during repair is shown in Figure 4.1 for a motorway viaduct repair on the crossbeam.

In some cases the need for structural support for live and dead loads and the effort to transfer loads into the repair (for instance as done as an exercise on the Midland Links elevated motorway sections) have made such repairs completely uneconomic. This is one reason for choosing an electrochemical rehabilitation technique such as cathodic protection, as only 'cosmetic' repairs are required and the minimum of concrete is removed.

6.3 Coatings, sealers, membranes and barriers

One of the attractions of construction in concrete is that coatings are not usually required so the maintenance costs are lower. In some countries there is a reluctance to coat concrete structures, in other counties it is more accepted. However, coatings can be beneficial in excluding undesirable species such as chlorides and carbon dioxide, or cosmetically restoring the appearance after concrete repair. An example of a coating to 'blend in' repairs to the original concrete is shown in Figure 6.6. Even the most carefully matched repair will weather differently from the original material, eventually requiring a coating to hide it again. There are architectural coatings designed to hide variations in the concrete appearance. While designed to deal with new construction the author has used them successfully on repaired structures. One particular formulation also has anticarbonation properties.

A huge range of coatings and sealers can be applied to concrete. Apart from looking different from each other, they also perform different tasks. There are anticarbonation coatings which should be applied after carbonation repairs to stop further carbon dioxide ingress. The carbonation resistance should always be checked rather than relying upon coating formulation as testing has shown that not all acrylics (for instance) have good carbonation resistance (Robinson, 1986). However, with the subsequent development of standards for carbon dioxide penetration of coatings, the quality and consistency of anticarbonation coatings has improved significantly. Penetrating sealers, silanes, siloxanes, siloxysilanes, etc. are recommended to reduce chloride ingress. They have been known to accelerate carbonation

Figure 6.6 (a) Roof structures before treatment showing obtrusive anticarbonation coatings and exposed reinforcement due to chloride and carbonation-induced corrosion. (b) Roof structures after repair and coating where the visual effect of patch repairs are minimized by application of Keim Concretal Lazure coating.

in laboratory tests so care should be taken in deciding where and when they are applied.

Once chloride-induced corrosion has started it is very unlikely that a coating of any sort will stop corrosion. There is sufficient moisture and oxygen within the concrete to generate the small amount of rust needed to crack the concrete and then the water and oxygen will have a free path to the steel surface. It is also unlikely that a coating applied in the field to a real structure will seal it totally. When chlorides get into concrete they form hygroscopic salts. These will condense water from the atmosphere and elsewhere in the concrete. It should be noted that the 'average' humidity in the United Kingdom (and much of the US East Coast) is around 70–80%. This is close to the optimum for corrosion. Any sealing coat is likely to seal that optimum humidity level into the concrete. It has now been demonstrated experimentally that coatings can control corrosion in carbonated structures but not where corrosion is due to chlorides (Sergi *et al.*, 2000).

The total or near total sealing of concrete with impermeable coatings can cause problems. Concrete is porous and contains water and oxygen in its pores. The amount of water as vapour and as liquid in the pores is in equilibrium with the atmosphere. If a coating seals water in the pores then when the atmospheric humidity drops or the temperature increases very large forces can be exerted against the barrier coating causing blistering and coating failure.

6.3.1 Carbonation repairs

If a combination of patching and sealing is required for a carbonated structure, the patch repairs must remove all carbonated concrete around the steel. In some cases this will only require exposure of the front face but in more extreme conditions it may need concrete removal behind the steel to restore the passive alkaline environment. The development of the standard for carbonation testing, BS EN 1062-6 Paints and varnishes – Coating materials and coating systems for exterior masonry and concrete – Part 6: determination of carbon dioxide permeability, means that selecting an anti-carbonation coating is now far more straightforward. The principle of anticarbonation coatings is that they are porous enough to let water vapour move in and out of the concrete but the pores are too small for the large carbon dioxide molecule to pass through. Anticarbonation coatings are generally high build coatings, with the carbonation resistance increasing with thickness. Work at the UK Building Research Establishment suggests that at least 200 μm is required to achieve adequate film thickness over the whole surface (Matthews *et al.*, 2003). Society Technical Report 50 (Concrete Society, 1997) discusses the different tests available for CO_2 diffusion resistance of coatings and the development of BS EN 1062-6 (BSI, 2002) giving test methods for anticarbonation coatings.

6.3.2 Coatings against chlorides, penetrating sealers

Penetrating sealers have been recommended as a way of stopping chlorides getting into concrete. The chemistry of the process is that silanes, siloxysilanes and similar chemicals will penetrate the pores of the concrete and react with the water in the pores to from a hydrophobic (water repelling) layer that stops water getting in as a liquid (that may carry salt with it), but allows water vapour in and out of the concrete so that it will 'breathe'.

The penetrating sealer is within the concrete so it is protected from physical damage and degradation by ultraviolet light, etc. The problem is that the colourless liquid must be applied so that most of it gets into the pores and there is enough water in the pores to react, but not too so much that it washes the sealer back out and pores remain uncoated. There is still some debate over the true penetration depth of sealers to good quality concrete in the field.

For several years from 1986 to the development of the European standards on coatings pure (100%) iso-butyl-trimethoxysilane has been specified to prevent chloride ingress on exposed concrete on UK highway bridges (Highway Agency, 2003). Present requirements are given in Highways England CD 373 (2020). A penetration depth of 2–4 mm has been claimed, but this may reduce to 1 mm or less in new, well cured concrete with a low water/cement ratio. Other highway agencies have used different formulation with some siloxanes in solutions and larger molecules that are less volatile. All these chemicals form silicone resins within the pores. The larger molecules are less volatile and therefore easier to get to the concrete surface, but once there they are less mobile and penetrating.

In the United States many highway agencies have applied penetrating sealers to concrete bridge decks. There has been no definitive research to show whether sealers work on trafficked surface or for how long they are effective.

The penetrating sealers approved for application to highway structures in Europe are generally solvent free and are applied in a series of coatings by thorough wetting of the surface with the sealer with a drying period between coats.

In the United States the SHRP program developed two field tests for penetrating sealers (Whiting et al., 1992). These can be used in situ to determine the effectiveness of sealers. One is a surface conductivity test and the other is a variation on the initial surface absorption test (ISAT), see Section 4.13. If silane-based sealers are being used then these tests are recommended. They can be used to rank the performance of different materials on a given structure. Once the structure is coated they will measure the effectiveness of the application and check uniformity across the structure.

The alternative is to take cores and do laboratory tests of chloride permeability. This is slow, expensive and damaging to the structure. There is now a European Standard BS EN 1062-3 for water transmission through coatings for concrete. See Section 4.13.

6.3.3 Waterproofing membranes

According to a report by the Organization for Economic Cooperation and Development (1989), most European countries put waterproofing membranes on their decks where there is a risk of chloride ingress. This is usually a sheet or a spray or 'squeegee' applied liquid system applied over the new concrete surface, sometimes with a base or primer coat and with protective layers, with a final asphalt overlay. A synthesis of practice of waterproofing membranes in North America was published by Manning (1995). Approximately 50 systems were tested by the UK Transport Research Laboratory in a trial of membrane systems (Price, 1989). Figure 6.7 is a schematic of the components of a waterproofing system.

These systems have been of variable quality in the past, although tighter specifications have been introduced by most national DOTs. Membranes have failed at joints, curbs and drains where chloride laden water could get under them. Some were damaged or destroyed by the application of the asphalt wearing course over them. An example is shown in Figure 6.8. A survey of UK Highway site practice and failures was given in Price (1991). Highways England requirements for the application of penetrating sealers is given in the Design Manual for Roads and Bridges CD 373 Impregnation of reinforced and prestressed concrete highway structures.

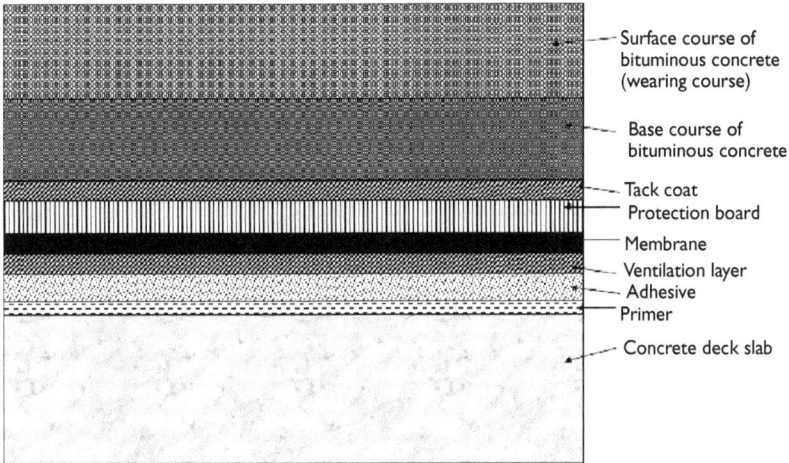

Figure 6.7 Schematic of possible components of a waterproofing system from Manning (1995).

Figure 6.8 Waterproofing membranes can fail to bond, dissolve into the concrete under application of the hot mix wearing course or be punctured by the aggregates. The photograph shows a failed membrane after testing at the UK Transport Research Laboratory.

However, the lack of waterproof membranes is why there are many cathodic protection systems on bridge decks in Canada and the northern states of the United States, where membranes were not widely used, while many European systems applied to bridges are on the support structure. Membranes and cathodic protection are not easily compatible as gases are evolved by CP systems which could be trapped by a waterproofing system. This is discussed in Chapter 7.

One problem with waterproofing membranes is that they have a 10- to 15-year life span. This means they must be replaced and any areas of concrete damage repaired. There is also a problem with severe pitting and 'black rust' on the reinforcing steel as discussed in Chapter 2 and as shown in Figure 6.9.

Waterproofing membranes have been developed for car park decks. These are generally very thin to minimize load and reduction in headroom. The main problem is 'scuffing' and 'shoving' from the tight turns and frequent braking of cars.

6.3.4 Barriers and deflection systems

These are the logical extension of the waterproofing membrane. Often one of the simplest ways of reducing the deterioration rate due to chloride attack is simple deflection of chloride laden water away from the concrete surface. This can sometimes be done with the introduction of guttering and drainage on buildings or bridge substructures subject to salt water run off. This can be a very cost-effective way of at least stopping the acceleration of decay. It can also extend the life of rehabilitation systems such as patches, cathodic protection anodes and chloride removal treatments.

A more expensive barrier approach can be seen on bridges where the reinforced concrete has been clad in a brick or masonry finish, usually for cosmetic reasons. The salt rarely penetrates to the concrete surface. However, this option is rarely available for reasons of cost. Encasement of

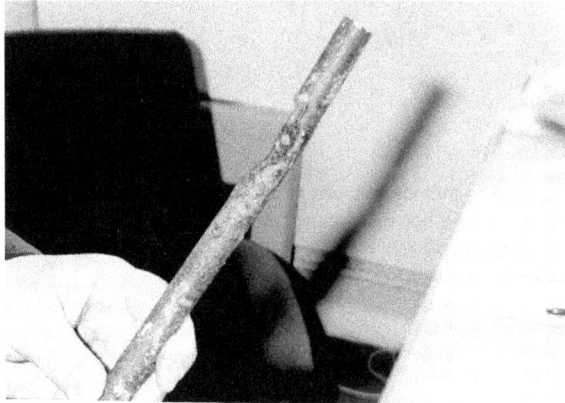

Figure 6.9 When membranes fail to bind to the concrete and water and chlorides get underneath, rapid local corrosion can lead to severe section loss without cracking or spalling of the concrete as the expansive corrosion product stays in solution.

building elements can often be undertaken as part of a refurbishment programme, such as with a change of use. External cladding can be added which encloses concrete elements at risk of corrosion. Once in a warm, dry, indoor environment the risk of corrosion is significantly reduced.

This approach is far less effective once the critical chloride threshold for corrosion has been exceeded. It is possible to determine the present chloride profile in the concrete and predict the likely build up of chloride at the reinforcement depth. This approach is discussed in Chapter 8.

6.4 Encasement and overlays

After milling, and deeper concrete removal where cracking and spalling has occurred, the deck is then patched in the delaminated areas and a dense cementitious overlay of microsilica, polymer modified or low slump, low water/cement ratio concrete is put back on. This will slow the corrosion rate and the appearance of further delaminations.

These techniques are popular in the United States for delaying, or preventing chloride induced delamination on substructures and bare concrete bridge decks. A review of highway bridge deck maintenance practice in the United States (Russell, 2004) found overlaying to be far more popular than the application of cathodic protection.

In marine conditions Florida DOT has found that encasing bridge columns in concrete is not effective in stopping corrosion. Prestressing strands in a marine pile were found to have corroded through shortly (three years)

Figure 6.10 Pre-tensioned tendons corroded through three years after encasement from Rapa and Hartt (1999). Courtesy Florida Department of Transportation.

after repair. The tendons were 7/16 inch in diameter. This implies a rate of over 3 mm/year if a 7/16″ = 11 mm tendon loses its full section in three years. This demonstrates that corrosion rates in concrete can be very high if encasement is used inappropriately (see Figure 6.10).

There was no indication of any problem with the reinforcement until the encasement was removed and it was found that there was steel left to restrain any sideways movement in the piles.

However, several DOTs in northern states are having some success with encasement. There are also trials of carbon fibre wraps used for strengthening as a method of suppressing corrosion.

Concrete overlays or encasement may absorb some of the chlorides, reducing the level at the concrete surface. They will certainly reduce the high chloride gradient that drives chloride further into the concrete if they are coupled with removal of some or all of the old cover concrete. This has been measured on bridge decks.

Recent research into overlays shows that many of them suffer from shrinkage cracks but still seem to be effective in stopping or slowing corrosion. Overlays last longest in the more southerly states like Virginia when compared with the more northerly ones like New York with colder winters, higher salt application and also higher traffic density.

Overlays may be of polymer modified concretes, low slump dense concrete (the Iowa mixes), or microsilica concretes. They vary in cost and ease of application. State highway agencies generally develop one or two mix designs that suit their purposes and make competitive bidding possible.

Encasement on substructure columns is less routine. The concrete is broken out where damage and an oversized shutter applied. Concrete is then pumped or placed into the shutter either enlarging the whole column or the damaged section.

6.5   Sprayed concrete

Sprayed concrete, shotcrete (dry sprayed) or gunite (wet sprayed) are methods of rapidly applying concrete to soffits or vertical surfaces. It can be used over patches and to overlay metal mesh anodes for cathodic protection as described in Section 7.4.4. It is sometimes applied as a temporary 'cosmetic' repair in the Northern USA and Canada when concrete has spalled or is in danger of spalling. As it does nothing to slow the corrosion rate it is comparable in effectiveness to patch repairing, and may suffer from the same 'incipient anode' problem.

When the concrete is applied as shotcrete, water and dry mix cement and aggregates are sprayed through two nozzles and mix 'in flight' and on impact. Considerable operator skill is required to apply shotcrete effectively with a minimum of delaminations. This is particularly true when cathodic protection expanded metal mesh anodes are being installed. The shotcrete is required to stick to the original surface which may not be as rough or absorptive as the surface exposed by concrete removal from corrosion damaged areas. Inadequate mixing or inappropriate quantities at the nozzle can lead to voids, honeycombing or lenses of unhydrated concrete. Extensive trials should be undertaken to show that good bond and mixing is achieved on site. Thorough inspection afterwards should show a high consistency of application achieved and maintained throughout the application.

Wet mix spraying of mortars is becoming more popular as proprietary mixes have been developed. The consistency of the mix is easier to control in wet mix application if segregation does not occur in the mixer before spraying. Application rates are higher and it is often possible to finish the surface of a wet sprayed mix. This is far more difficult with dry mixed shotcrete.

6.6   Corrosion inhibitors

Corrosion inhibitors for steel in concrete are of great interest to the concrete repair community. In this section we will discuss what they are, how they work and what work is going on to assess their effectiveness.

There are various definitions of a corrosion inhibitor. The definition is important because many things can inhibit corrosion, such as coatings, a vacuum or cathodic protection. A definition is provided in ISO 8044 1989: 'A chemical substance that decreases the corrosion rate when present in the corrosion system at suitable concentration, without significantly changing the concentration of any other corrosion agent.' This definition excludes coatings, pore blockers and other materials that act on the water, oxygen and chloride concentrations and clarifies what we are talking about.

In the case of steel in concrete we are talking about chemicals that are either admixed into fresh concrete or that transport through the hardened concrete to react on the reinforcing steel surface to slow down the rate of corrosion. This difference is fundamental. By admixing during the construction process the precise dosage is controlled and conditions can be readily tested. Applying inhibitors to hardened concrete as part of a rehabilitation system is far more problematical in ensuring effective dosage at the rebar, and in measuring, testing and predicting long-term performance.

There are a number of ways of subdividing inhibitors:

A. By their action:
The corrosion reaction in concrete occurs by the formation of anodes and cathodes (Figure 2.2). Corrosion inhibitors can therefore be:

- Anodic inhibitors – suppressing the anodic corrosion reaction
- Cathodic inhibitors – suppressing the cathodic reaction
- Ambiodic inhibitors – suppressing both anodes and cathodes.

B. By their chemistry and function:

- Inorganic inhibitors – nitrites, phosphates and other inorganic chemicals
- Organic inhibitors – amines and other organic chemicals
- Vapour phase or volatile inhibitors – a subgroup of the organic inhibitors (generally aminoalcohols) that have a high vapour pressure.

All of these inhibitors have been widely used by corrosion engineers for many years to protect steel and other metals such as electronic components. In some cases their chemistry is well understood, in other cases less so.

6.6.1 Corrosion inhibitors for admixing into fresh concrete

Specifically for steel in concrete there are several types of inhibitors. There are those that are admixed into new concrete and those that are applied to hardened concrete as part of a rehabilitation programme of a corrosion damaged structure. Of the available materials the most widely applied is a proprietary calcium nitrite admixture to new concrete which is of proven effectiveness as long as the chloride : nitrite ratio stays less than about 1.8:1 (Virmani and Clemena, 1998). A dosage rate of 10–30 litres per cubic metre of concrete is generally specified, depending on the expected maximum chloride level at the rebar. This has proven efficacy and is usually the bench-mark against which other inhibitors are tested. Competitor companies offer other proprietary materials. They are generally organic inhibitors with active amine and ester groups. The main problem with most organic and inorganic inhibitors is that they act as severe set retarders. These formulations of amines, esters and nitrites do not. In fact nitrites act as set accelerators (their original use as a concrete additive was as a non-chloride set accelerators for use in cold climates). In most conditions nitrites are sold with a set retarder in the formulation to avoid flash setting.

The merits of admixed corrosion inhibitors have been debated for many years. They are consumed as they compete with chloride ions to block the anodic sites (where they are anodic inhibitors). However, the excess inhibitor available in the concrete should maintain a constant availability at the steel surface for any reasonable lifetime. There is increasing evidence of their stability although there has been some concern about the long-term leaching of the inhibitor, particularly in concrete exposed to a marine environments (Virmani and Clemena, 1998). There are several structures several decades old protected by nitrite inhibitors. There are at least half-a-dozen marine structures in the United Kingdom at least partly protected from reinforce-ment corrosion by calcium nitrite. The organic admixtures are newer and have less of a track record both in the field and in long-term testing.

The upsurge in interest in admixed inhibitors came from the failures of fusion bonded epoxy coated reinforcing steel, mainly in the Florida Keys Bridges in the Gulf of Mexico in the late 1980s. Epoxy coated steel had become the protection system of choice in the United States for bridge decks and substructures. When the failures were reported, the Florida Department of Transportation and the Federal Highways Administration were flooded with offers of alternative protection systems. Although Florida no longer uses epoxy coated reinforcing steel, most state and provincial highway agencies in North America still find it satisfactory, and cheaper than the alternatives. Florida DOT are assessing the long-term effectiveness of inhibitors for marine bridge applications. Proprietary formulations are offered more widely in North America than in Europe. A review of corrosion inhibiting admixtures has recently been published by AMPP (2018). The European Federation of Corrosion published a state-of-the-art Report 35 on inhibitors for steel in concrete, this focuses mainly on inhibitor admixtures but does discuss topically applied products. Elsener (2001) and The Corrosion Prevention Association have published Technical Note 16 which discusses inhibitors applied to hardened concrete (Atkins *et al.*, 2016).

6.6.2 Corrosion inhibitors for hardened concrete

Vapour phase inhibitors or migrating corrosion inhibitors have been used to impregnate packaging, greases and waxes for many years to protect steel machinery and components, particularly before use. An American company realized in the 1980s that they might be effective in diffusing through concrete pores and protecting reinforcing steel. Several companies now offer proprietary formulations, which are again amine and ester based, with amino alcohols as the main volatile component.

Amino alcohols are ambiodic, forming a film on the steel surface, blocking both anodic and cathodic reactions. They can be applied as a coating on the surface of the concrete, a 'plug' of material in a hole, or admixed into repairs. The suppliers claim that these materials will move very rapidly through the air voids in the concrete, through pores and micro-cracks to reach the steel and protect it.

A migrating inhibitor with no pretensions of volatility is MFP or monofluorophosphate. This relies on capillary action of other transport mechanisms to get the material down to the steel through the concrete cover. A recent paper showed that it could be quite successful in carbonated concrete (Raharinaivo and Malric, 1998). There is discussion of migrating inhibitors in Elsener (2001) and Atkins *et al.* (2016).

6.6.3 Track record, case histories and monitoring

One issue with corrosion inhibitors is to determine how effective they are. There are two aspects to this: Can they be detected and levels measured quantitatively in the concrete and at the rebar surface? Are they effectively suppressing the corrosion rate?

Detection is obviously specific to a particular active inhibiting chemical. Some tests are expensive and difficult. Much work has gone on to develop a test for aminoalcohols, with mixed results so far as this author is aware. As stated earlier, it is far easier to carry out tests on admixed inhibitors than for those applied to hardened concrete. There can be dangers with under dosing inhibitors, in that corrosion may occur very aggressively in small anodic pits. It is therefore important that the risk of pitting and under dosing is fully understood.

The most effective method of assessing corrosion inhibitors is to measure the corrosion rate. If a corrosion inhibitor is effective it should reduce the active corrosion rate by at least one and preferably two orders of magnitude. The author has reported long-term monitoring of structures which had corrosion inhibitors applied in 1995 (Broomfield, 1997). Aminoalcohol-based inhibitors were applied by surface application, down holes drilled in the concrete and in repair patches to support structures with over 1% chloride by weight of cement cast into the concrete as calcium chloride set accelerator. Over the years a large number of patch repairs have been carried out. This trial on over 60 structures showed only a slight reduction in corrosion rate and the onset of cracking of treated structures at approximately the same time as untreated controls. In a very small scale trial with another proprietary aminoalcohol-based inhibitor, short-term results were more encouraging, with a 91% reduction in the corrosion rate measured with a linear polarization device in an area showing active corrosion rates and corrosion potentials. However, there has been no follow up to this trial on a car park deck with low cover, which would maximize the likelihood of inhibitor reaching the steel.

The UK Transport Research Laboratory set up laboratory tests on a range of inhibitors available in the United Kingdom. They are also doing field tests on bridge bents on the Midland Links Motorway near Birmingham for the Highways Agency. The UK Building Research Establishment (BRE) has had long-term outdoor exposure trials going for several years. BRE also did tests many years ago on calcium nitrite (Treadaway and Russell, 1968).

In the United States corrosion inhibitor testing goes back many years. The calcium nitrite inhibitor cast into concrete has been widely tested in the laboratory and the field. Aminoalcohol-based inhibitors were first trialed in hardened concrete in the 1980s, with some of the most comprehensive testing done by the Strategic Highway Research Program (Al-Qadi et al., 1993; Prowell et al., 1993). A recent review of the field trials on bridge decks suffering from chloride ingress showed no beneficial effect of inhibitor treatments (Sohanghpurwalla et al., 1997). Similarly, a large scale exposure trial of inhibitors on slabs also found no beneficial effects of a number of inhibitors (Sprinkel and Ozyildirim, 1998). The exception was calcium nitrite cast into the concrete.

If coatings are applied after the inhibitor is introduced then corrosion monitoring is made harder. In order to take measurements of the corrosion condition, either the coating must be removed locally or (preferably) corrosion rate probes embedded in the concrete, as has been done on some car parks in the United Kingdom (Broomfield *et al.*, 1999). Ideally this should be done before the inhibitor is applied so that 'control' measurements can be taken prior to treatment. Permanent monitoring probes will effectively monitor the changes in corrosion rate without disturbing the structure, and ensuring that readings are made in consistent ways in consistent locations. It should be noted when using embedded reference electrodes that there is the potential for inhibitors to migrate into the reference electrode and either changing the measured potential or inhibiting the half cell reaction, shutting down the reference electrode.

6.6.4   Summary of the use of corrosion inhibitors in repairs

This section cannot provide a definitive review or state-of the-art review of corrosion inhibitors for concrete. It looks at the various issues facing those who want to expand the use of corrosion inhibitors. The biggest issues are for those wishing to use them for rehabilitation rather than for durable construction.

The research and field trials on corrosion inhibitors reveal the following issues:

- There is very little field data on corrosion inhibitors.
- The available field data there is often poor, with no clear evidence of the amount of inhibitor applied, whether it reached the rebar and if it is reducing corrosion rates and extending time to cracking.
- Many claims have been made about the transport of inhibitors through hardened concrete, these need to be independently assessed.
- We will need definitive evidence of the dosage vs. chloride level to achieve a given (low) corrosion rate.
- For application to hardened concrete we need quantitative data on its penetration vs. concrete cover and concrete permeability.
- We need more information on the performance of inhibitors, particularly well controlled field trials, and long-term corrosion monitoring.

Corrosion inhibitors are inexpensive materials that can be applied simply. However, until there is clear evidence of their envelope of effectiveness in terms of chloride level, corrosion rate, dosage cover and longevity, it will be difficult to do comparative whole life costing of inhibitor application vs. proven alternatives such as cathodic protection.

This is particularly so for inhibitors for hardened concrete. In the case of inhibitors for admixing into fresh concrete, whole life costing programmes are available from the manufacturers that show the cost effectiveness of their inhibitor vs. some of their other products or alternative methods of improving durability in aggressive environments.

It may prove that inhibitors for hardened concrete are most effective for carbonated and low chloride level concretes, with low cover and reasonable permeability. However, laboratory research found little effect of organic inhibitors on the corrosion of steel in carbonated concrete (Bolzoni *et al.*, 2006). They may not be suitable for rehabilitation of highway structures with high chloride levels and high strength structural concrete with low permeability. While some of the published and unpublished field data support this view, there has been few published recommendations giving the range of conditions under which they are effective. However, there is much research both ongoing and reported, by manufacturers and independent researchers, for example under the COST 521 programme in Europe and the Florida DOT and FHWA funded research in the United States.

At this point the author would consider any use of the proven corrosion inhibiting admixtures for new construction to be an effective corrosion prevention measure if backed up with proper assessment and whole life costing. However, any application of inhibitors for repair of existing structures should be done on a trial basis with proper long-term monitoring of the treatment. A review of inhibitors admixed and applied to hardened concrete has been published by Atkins, Merola and Foster (2016).

6.7 Standards and guidance on physical and chemical repair

However a concrete repair is carried out, by hand placing mortar, sprayed or pumped into a new form, the important factor is that it is well consolidated and well adhered. To this end a coring machine should be used to core through repairs to the substrate, a pull off test done and the core evaluated for its consolidation and compressive strength. Very few standards or guidance documents are willing to give definitive values for the pull off tests on repairs in the field. One of the few is Concrete Society (1991) which states that the minimum pull of strength should be at least 0.8 N/mm² unless failure occurs within the parent concrete. EN1504-10 (2003) states in its non-mandatory appendix Table A.2 that 'Site values within the range 1.2–1.5 MPa for structural repairs and a minimum of 0.7 MPa for non structural repair are acceptable'. Values for laboratory performance are given in EN1504-3.

There is a systematic European standard to physical and chemical techniques for concrete repair of chloride contaminated structures. This is BS EN 1504-9. Products and repair Systems for the Protection and repair of Concrete Structures – Definitions, requirements, quality control and evaluation of conformity. Part 9 General Principles for the use of products and systems. The earlier parts cover different classes of materials while part 10 covers site application:

Part 1 – Definitions
Part 2 – Surface protection systems for concrete
Part 3 – Structural and non-structural repair
Part 4 – Structural bonding
Part 5 – Crack injection
Part 6 – Anchoring of reinforcing steel bar
Part 7 – Reinforcement protection
Part 8 – Quality control and evaluation of conformity
Part 9 – General principles for the use of products and systems
Part 10 – Site application of products and systems and quality control of the works.

CE Marking is now well established for the various concrete repair and protection products covered in parts 2 to 7 using the methodology in Part 8. Part 9 gives principles of repair depending on whether the damage is caused by corrosion or not. Part 10 guides designers in the development of specifications and guides applicators in the application of the works. Part 10 has a very useful informative (non-mandatory) appendix. The use of EN1504 will be discussed further in Chapter 8 on rehabilitation methodology. The UK remains a member of CEN, the umbrella body for European Standards, however, CE marking of construction products sold in the UK will be replaced by UKCA (UK Conformity Assessed) marking in 2023 although products marketed in the European Union will of course still require CE marking as well as UKCA. This should mean that the only difference UK purchasers of Construction products should notice is the addition of the UKCA mark.

A comprehensive Concrete Repair Manual has been developed by ACI in collaboration with BRE, ICRI and the Concrete Society. This was last revised in 2013 (ACI, 2013). This includes the ACI, Concrete Society, BRE and ICRI guides on condition evaluation and diagnosis, patch repairing and overlaying, as well as guidance documents on protection and corrosion management. Concrete Society (1991) is part of the manual. The manual also includes Corrosion Prevention Association (CPA) monographs on electrochemical techniques, and a limited selection of NACE/AMPP recommended practice or technical reports.

National Highway Agencies will also have internal standards and guidance for concrete repairs. In the UK Highways England, and its sister bodies in Wales, Scotland and Northern Ireland, have published a Design Manual for Roads and Bridges, which includes CS 462 Repair and Management of deteriorated concrete highway structures (HE 2020-1). There is also a

Manual of Contract Documents of Highway Works which includes standard specification for concrete repair (HE 2020-2).

References

ACI (2013). Concrete Repair Manual. 2nd Edition. 2013; Two Vols Publ. ACI International, BRE, Concrete Society, International Concrete Repairs Institute.

Al-Qadi, I.L., Prowell, B.D., Weyers, R.E., Dutta, T., Gouru, H. and Berke, N. (1993). Concrete Bridge Protection and Rehabilitation: Chemical and Physical Techniques – Corrosion Inhibitors and Polymers SHRP-S-666, Strategic Highway Research Program, National Research Council, Washington, DC.

AMP (2018). Corrosion Inhibiting Admixtures for Reinforced Concrete – A State of the Art Review TR 21428, Association for Materials Performance and Protection, Houston, TX.

Anon. (1993). 'Hydro-demolition'. *Construction Repair*, 7(2): 32–36.

Atkins, C., Merola, R. and Foster, A. (2016). Corrosion Inhibitors, Technical Note 16, Corrosion Prevention Association, Bordon, UK.

Bolzoni, F., Lazzari, L., Ormellese, M. and Goidanich, S. (2006). Prevention of Corrosion in Concrete the Use of Admixed Inhibitors, Proc. Corrosion 2006, San Diego CA, NACE International Houston.

Broomfield, J., Davis, K. and Hladky, K. (April 1999). Permanent Corrosion Monitoring for Reinforced Concrete Structures. Paper 99559, Corrosion/99, NACE International, Houston, TX.

Broomfield, J.P. (1997). 'The Pros and Cons of Corrosion Inhibitors'. *Construction Repair Journal*, 11(4): 16–18.

BSI (1999). Paints and Varnishes – Coating Materials and Coating Systems for Exterior Masonry and Concrete Part 3 – Determination and Classification of Liquid Water Transmission Rate (permeability) EN 1062–3, British Standards Institute, London.

BSI (2002). Paints and Varnishes – Coating Materials and Coating Systems for Exterior Masonry and Concrete Part 6 – Determination of Carbon Dioxide Permeability BS EN 1062–6, British Standards Institute, London.

CD 373 Impregnation of Reinforced and Prestressed Concrete Highway Structures, Highways England, Design Manual for Roads and bridges.

CIRIA (1993). Standard Tests for Repair Materials and Coatings for Concrete Part 3: Stability, Substrate Compatibility and Shrinkage Technical Note 141, CIRIA, London, UK.

Concrete Society (1991). Patch Repair of Reinforced Concrete – Subject to Reinforcement Corrosion. Technical Report No 38. The Concrete Society, Camberley, UK.

Concrete Society (1997). Guide to Surface Treatments for Protection and Enhancement of Concrete. Technical Report 50. The Concrete Society, Camberley, UK.

Elsener (2001). Corrosion Inhibitors for Steel in Concrete, State of the Art Report European Federation of Corrosion Publication 35, Institute of Materials, London.

Emberson, N.K. and Mays, G.C. (1990). Design of Patch Repairs: Measurements of Physical and Mechanical Properties of Repair Systems for Satisfactory Structural Performance. In Protection of Concrete, R.K. Dhir and J.W. Green (eds), E&FN Spon, London, UK.

ICRI (2004). Guideline for the Preparation of Concrete Surfaces for Repair Using Hydrodemolition Methods Technical Guide G03737 International Concrete Repair Institute Des Plaines, IL, USA.

HE (2020–1). CS462 Repair and Management of Deteriorated Concrete Highway Structures, Highways England.

HE (2020–2). Manual of Contract Documents for Highway Works Volume 1 Specification for Highway Works Series 7500 Concrete Repairs, Highways England.

Ingvarsson, H. and Eriksson, B. (1988). 'Hydrodemolition for Bridge Repairs'. *Nordisk Betong*, 2–3: 49–54.

Manning, D.G. (1995). Waterproofing Membranes for Concrete Bridge Decks: A Synthesis of Highway Practice, NCHRP Synthesis 220, National Cooperative Highway Research Program, Transportation Research Board, National Research Council, Washington, DC.

Matthews, S., Murray, M., Boxall, J., Bassi, R. and Morelidge, J. (2003). Maintenance of Concrete Buildings and Structures. In Concrete Building Pathology, Susan Macdonald (ed.), Blackwells, Oxford, UK.

Organization for Economic Cooperation and Development (1989). Durability of Concrete Road Bridges, OECD, Paris.

Price, A.R.C. (1989). A Field Trial of Waterproofing Systems for Concrete BridgeDecks, TRRL Research Report 185, Transport Research Laboratory, Crowthorne, Berkshire, UK.

Price, A.R.C. (1991). Waterproofing of Concrete Bridge Decks: Site Practice and Failures, TRRL Research Report 317, Transport Research Laboratory, Crowthorne, Berkshire, UK.

Prowell, B.D., Weyers, R.E. and Al-Qadi, I.L. (1993). Concrete Bridge Protection and Rehabilitation: Chemical and Physical Techniques – Field Validation SHRP-S-658, Strategic Highway Research Program, National Research Council, Washington, DC.

Raharinaivo, A. and Malric, B. (December, 1998). Performance of Monofluorophosphate for Inhibiting Corrosion of Steel in Reinforced Concrete Structures, Proc. International Conference on Corrosion and Rehabilitation of Reinforced Concrete Structures, Orlando, FL.

Rapa, M. and Hartt, W.H. (1999). Non-Destructive Evaluation Of Jacketed Prestressed Concrete Piles For Corrosion Damage. NACE Corrosion 99. (Paper No 566), NACE International, Houston, TX, USA.

Robinson, H.L. (1986). Evaluation of Coatings as Carbonation Barriers. Proceedings of the Second International Colloquium on Materials Science and Restoration, Technische Akademie, Essingen, Germany.

Russell, H.G. (2004). Concrete Bridge Deck Performance: A Synthesis of Highway Practice. National Cooperative Highway Research Program Synthesis 333, Transportation Research Board, Washington, DC, USA.

Sergi, G., Seneviratne, A.M.G., Maleki, M.T., Sadegzadeh, M. and Page, C.L. (2000). 'Control of Reinforcement Corrosion by Surface Treatment of Concrete'. *Proc. Instn. Civ. Engrs. Structures & Buildings*, 140(1): 85–100.

Sohanghpurwalla, A.A., Islam, M. and Scannell, W. (1997). Performance and Long Term Monitoring of Various Corrosion Protection Systems used in Reinforced Concrete Bridge Structures. Proceedings of International Conference Repair of Concrete Structures, from Theory to Practice in a Marine Environment.

Sprinkel, M. and Ozyildirim, C. (1998). Evaluation of Exposure Slabs Repaired with Corrosion Inhibitors. Proc. International Conference on Corrosion and Rehabilitation of Reinforced Concrete Structures, Orlando, FL, USA.

Treadaway, K.W.J. and Russell, A.D. (1968). 'Inhibition of the Corrosion of Steel in Concrete'. *Highways and Public Works*, 36(1704): 19–21.

Vassie, P.R. (1987). Durability of Concrete Repairs: The Effect of Steel Cleaning Procedures, Research Report 109, Transport Research Laboratory, Crowthorne, Berkshire, UK.

Virmani, Y.P. and Clemena, G.G. (1998). Corrosion Protection – Concrete Bridges. Federal Highways Administration Report FHWA-RD-98-088, Washington, DC, USA, p 30.

Vorster, M., Merrigan, J.P., Lewis, R.W. and Weyers, R.E. (1992). Techniques for Concrete Removal and Bar Cleaning on Bridge Rehabilitation Projects SHRP-S-336 National Research Council, Washington, DC, USA.

Weyers, R.E., Prowell, B.D., Sprinkel, M.M. and Vorster, M. (1993). Concrete Bridge Protection, Repair, and Rehabilitation Relative to Reinforcement Corrosion: A Methods Application Manual. SHRP Reports. SHRP-S-327.

Whiting, D., Ost, B. and Nagi, M. (1992). Condition Evaluation of Concrete Bridges Relative to Reinforcement Corrosion, Volume 5: Methods for Evaluating the Effectiveness of Penetrating Sealers. Strategic Highway Research Program Report SHRP-S-327, National Research Council, Washington, DC, USA.

Whiting, D., Todres, A., Nagi, M., Yu, T., Peshkin, D., Darter, M., Holm, J., Anderson, M. and Geiker, M. (1993). Synthesis of Current and Projected Concrete Highway Technology, Stratgic Highway Research Program Report SHRP-C-345, National Research Council, Washington, DC, USA.

Chapter 7

Electrochemical repair techniques

7.1 Basic principles of electrochemical techniques

As we saw in reactions (2.1) and (2.2), corrosion occurs by the movement of electrical charge from the anode (a positively charged area of steel where the steel is dissolving and forming rust) to the cathode (a negatively charged area of steel where a charge balancing reaction occurs, turning oxygen and water into hydroxyl ions). This means that the process is both electrical and chemical, that is, electrochemical. We have also seen that, in the case of chloride attack, patch repairs are only a local solution to corrosion and repairing an anode may accelerate corrosion in adjoining areas due to the incipient anode effect (Section 6.2.1 and Figures 6.4 and 6.5).

One solution to this problem is to apply an electrochemical treatment which will suppress corrosion across the whole of the treated structure, element or area at risk. Figure 7.1 shows the basic components of an electrochemical system. They are a variable DC power supply, a control and monitoring system, an anode (temporary or permanent) usually distributed across the surface either fixed to it or embedded in the cover concrete, and monitoring probes, usually in the form of embedded reference electrodes. Electrochemical methods work by applying the external anode and passing current from it to the steel so that all the steel is made into a cathode. Three techniques are described here. The best known and most established technique is cathodic protection. This subdivides into impressed current cathodic protection and galvanic (also known as sacrificial) cathodic protection. An alternative using a temporary anode and a short-term (typically 4 to 8 week) treatment time for chloride infested structures is known as electrochemical chloride extraction (also known as electrochemical chloride removal and as desalination). A similar 1 to 2 week treatment method for reinforcement subject to carbonation-induced corrosion concrete has also been developed. This is known as electrochemical realkalization. This chapter discusses these techniques, their advantages and limitations and their application.

All of these treatments are suitable for reinforcing steel in concrete. However, the passing of electric current and of the voltages used can have

DOI: 10.1201/9781003223016-7

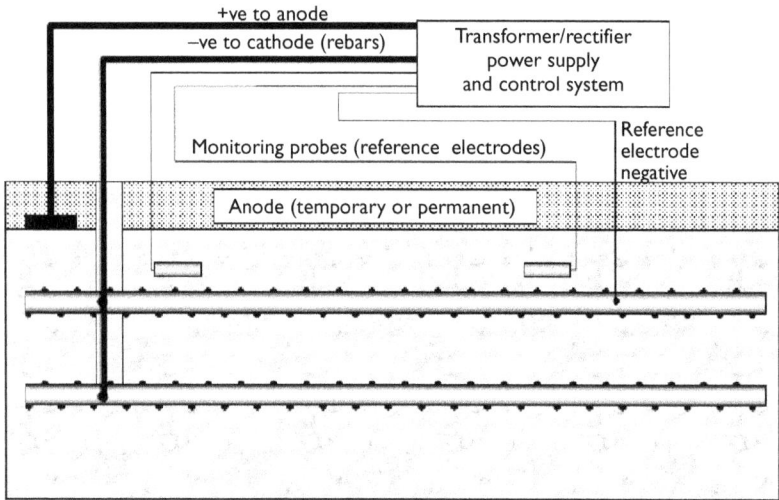

Figure 7.1 Schematic of electrochemical protection systems.

adverse effects on concrete and on steel. They can only be applied with great care to structures containing prestressing steel (pre- or post-tensioned) and to structures suffering from alkali-silica reactivity (ASR). They cannot be applied through electrically insulating layers or patches or through air voids. These issues are discussed later.

7.2 Cathodic protection principles

7.2.1 Theory and principles – impressed current systems

Impressed current cathodic protection works by passing a small direct current (DC) from a permanent anode on top of or fixed into the concrete through the concrete to the reinforcement. The power supply passes sufficient current from the anode to the reinforcing steel to force the anode reaction (7.1):

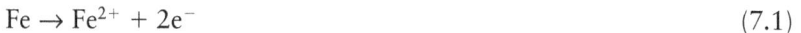

$$Fe \rightarrow Fe^{2+} + 2e^- \tag{7.1}$$

to stop and make cathodic reactions occur on the steel surface such as (7.2):

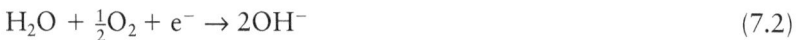

$$H_2O + \tfrac{1}{2}O_2 + e^- \rightarrow 2OH^- \tag{7.2}$$

The reinforcement network then becomes cathodic and corrosion is suppressed. However, another cathodic reaction can occur if the potential gets too negative:

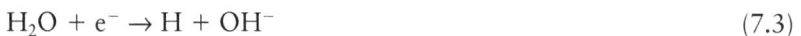

$$H_2O + e^- \rightarrow H + OH^- \tag{7.3}$$

This hydrogen evolution reaction can lead to hydrogen embrittlement. Monatomic hydrogen can diffuse into the steel and condense at grain boundaries or other defects in the crystalline matrix of the steel, weakening it and causing failure under load. This problem is negligible for normal reinforcing steel, but is of considerable concern for prestressed structures where the high tensile steel can be very susceptible to hydrogen embrittlement and the loading of the steel to up to 75% of its ultimate tensile strength can make it liable to catastrophic failure.

The problems of hydrogen embrittlement and of gas evolution are usually controlled by limiting the potential of the steel to below the hydrogen evolution potential. However, in acidic pits or crevices it may be possible for the potential to exceed the hydrogen evolution potential without it being measured by the potential against the reference electrodes. The cathodic protection of prestressed structures should only be undertaken with great care and input from experienced corrosion experts. An excellent state-of-the-art report on the cathodic protection of prestressed has been published (NACE, 2018).

The generation of hydroxyl ions in equations (7.2) and (7.3) will increase the alkalinity and help to rebuild the passive layer where it has been broken down by the chloride attack.

The chloride ion itself is negative and will be repelled by the negatively charged cathode (reinforcing steel). It will move towards the (new external) anode. With the carbon-based anodes it may then combine to form chlorine gas at the anode:

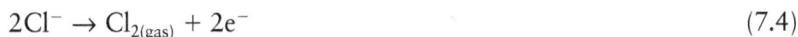

$$2Cl^- \rightarrow Cl_{2(gas)} + 2e^- \tag{7.4}$$

The other major reaction at all major anodes is the formation of oxygen:

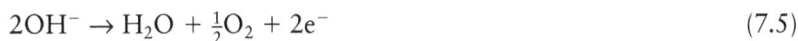

$$2OH^- \rightarrow H_2O + \tfrac{1}{2}O_2 + 2e^- \tag{7.5}$$

and

$$H_2O \rightarrow \tfrac{1}{2}O_2 + 2H^+ + 2e^- \tag{7.6}$$

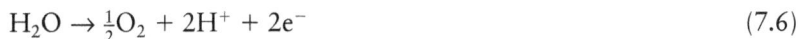

Reaction (7.5) is the reverse of reaction (7.2), that is, alkalinity is formed at the steel cathode (enhancing the passivity of the steel) and consumed at the anode. These and related reactions can acidify and carbonate the area around the anode (especially where carbon-based anodes are used, where the carbon also turns into carbon dioxide) and can lead to etching of the concrete surface and attack on the cement paste and even some aggregates once the alkalinity is consumed.

We can therefore see that three factors must be taken into account when controlling our cathodic protection system:

1   There must be sufficient current to overwhelm the anodic reactions and stop, or severely reduce the corrosion rate.
2   The current must stay as low as possible to minimize the acidification around the anode and the attack on the anode for those that are consumed by the anodic reactions.
3   The steel should not exceed the hydrogen evolution potential, especially for prestressed steel to avoid hydrogen embrittlement.

The balancing of these requirements will be discussed later under criteria for control of impressed current cathodic protection systems.

One of the more confusing facts of cathodic protection is that when we carry out a reference electrode potential survey of a reinforced concrete structure (Section 4.8) the most negative areas are those that are at highest risk of corrosion while the areas with a positive potential are at the lowest risk of corrosion, that is, cathodic (Section 4.8.3). However, to achieve cathodic protection, we must depress the potential of the steel. The reasons for this are explained later.

One of the earliest explanations of the criteria for achieving effective cathodic protection is to depress the potential of the cathodes to that of the most anodic areas (Mears and Brown, 1938). This stops the current flow from anode to cathode. It works because cathodes are more easily polarized than anodes. As discussed in Section 4.12 on corrosion rate measurement, for a fixed current, an actively corroding area shifts its potential less than a non corroding area. Therefore, once we depress all the cathodes below the potential of the anodes, corrosion stops.

Another way of looking at it is to consider the Pourbaix diagram Figure 7.2 (Pourbaix, 1973; Morgan, 1990) for iron in chloride solutions. This shows pH and potential changes as the iron moves from areas of corrosion to areas of passivity and then to immunity from corrosion.

Ideally, we would like to depress the potential sufficiently to reach the immune zone. However, that is very close to the hydrogen evolution potential (lower dotted line) at pH 12 which is where steel in concrete lies. For the reasons discussed earlier in this section, we want to avoid hydrogen evolution, so we aim for the area below the pitting potentials. This is discussed in Appendix 1 of BS EN ISO 12696 (2022) and also in Section 7.6.

7.2.2   The history of cathodic protection of steel in concrete

The principles of galvanic anode cathodic protection were discovered by Sir Humphrey Davy in 1824. His results were used over the next century or so to protect the submerged metallic parts of ships from corrosion. In the early decades of the 20th century the technology was applied to underground

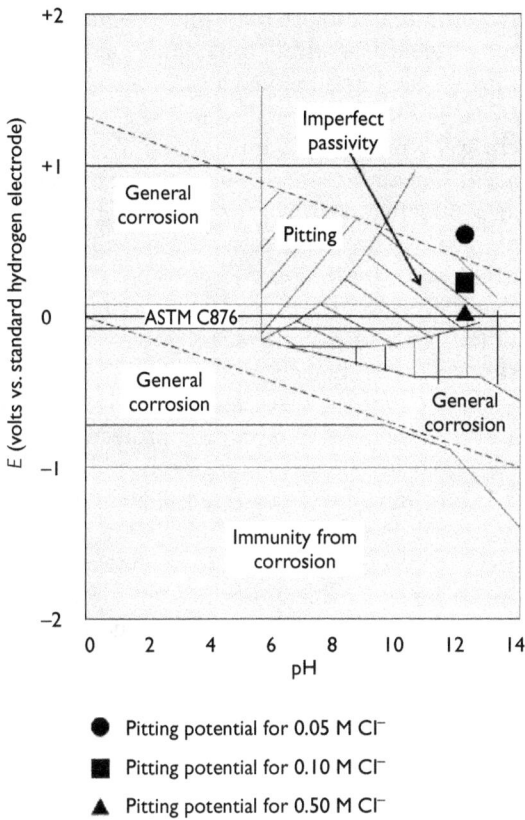

Figure 7.2 Pourbaix diagram showing experimental conditions of immunity to corrosion, general corrosion and imperfect passivity for iron in 355 ppm (0.01 M) chloride along with the pitting potentials at higher chloride levels.

pipelines. Impressed current cathodic protection was developed when it was found that the electrolytes like soils had too high electrical resistance for galvanic systems to be effective.

The problem of corrosion of steel in concrete was first ascribed to stray current flows from trams and DC railway systems (Hime, 1994). Once chloride, in the form of deicing salt, was identified as the major culprit (when chlorides took over from grit as a road deicer in the 1960s) an enterprising engineer in the California Department of Transportation (Caltrans) took a standard pipeline cathodic protection design and 'flattened it out' on a bridge deck. The system was straightforward. One of the popular impressed current pipeline cathodic protection anodes of that time was made of a corrosion resistant silicon iron, surrounded by a carbon coke breeze backfill. A well was dug near the pipeline, the anode put in surrounded by the backfill and the system connected to a DC power supply, with the negative terminal connected to the pipeline to make a cathodic protection system. Richard Stratfull took 'pancake' silicon iron anodes, fixed them on a bridge deck and applied a carbon coke breeze asphalt overlay (Stratfull, 1974). The systems installed in 1973 and 1974 were reviewed in 1989 and were still working (Broomfield and Tinnea, 1992). They have since been replaced. The author conducted some of the earliest research for the UK DoT on conductive coatings for cathodic protection of buildings (Geoghegan et al., 1985). The first trial in the United Kingdom was designed and installed by the author on Melbury House above Marylebone Station in London for British Rail in 1984. This included a remote control monitoring by a modem link (Broomfield et al., 1987).

Four trial systems were designed and installed in 1987 for the UK DoT on the Midland Links motorway system around Birmingham, Britain's second largest city. Those systems provided protection for over 20 years when some were replaced with titanium mesh and overlay systems and about 600 cross-beams have been protected (Unwin and Hall, 1993) with further beams protected on the Midland Links since then and older anodes and control systems replaced. An example is shown in Figure 4.1. Early systems were also designed by the author in 1986–1987 and were installed on bridges and buildings in Hong Kong and marine structures in Australia. There are now a large number of systems on bridges and other structures all around the world. In 2001, there was an estimate of 2–3 million m^2 of impressed current anode applied to structures worldwide. It has been estimated that 0.5 to 1 million sq m per annum of concrete is being protected with impressed current anode systems (Broomfield and Wyatt, 2018).

Since those first systems were applied in the 1970s, systems have been developed and applied to bridge decks, substructures and other elements, buildings, wharves and every conceivable type of reinforced concrete structure suffering from corrosion of the reinforcing steel. More recently systems have also been applied to steel in mortar in stone brick and terracotta clad structures on early 20th century steel framed historic buildings and structures (Chess and Broomfield, 2014).

Anodes have been developed in the form of conductive coatings, metals embedded in concrete overlays, conductive concrete overlays and probes drilled into the concrete. Anodes continue to be developed, applied in new configurations and to new structures. In the next section we will discuss the major components of cathodic protection systems, and particularly the anode systems that are available as these are the most prominent part of the cathodic protection system. Judicious choice of cathodic protection anode can maximize the cost effectiveness of the system.

The reason for choosing cathodic protection is almost always long-term cost effectiveness. The cathodic protection system prevents corrosion across the whole of the protected area of the structure, unlike localized patch repairs, particularly where chloride-induced corrosion is the problem. If repeated cycles of patch repairs are too expensive or unacceptable, then a properly maintained cathodic protection will usually work out as more cost effective in the long term. If the structure only has a short-term future and patch repairing is acceptable and inexpensive, then cathodic protection is not usually a suitable cost effective option.

7.3 Galvanic anode systems

7.3.1 Principles of galvanic cathodic protection systems

As discussed earlier, there are two forms of cathodic protection, impressed current and galvanic. The impressed current system has been described earlier and is the system with the longest history of application to atmospherically exposed reinforced concrete structures. An alternative method is to connect the steel to a sacrificial or galvanic anode such as zinc. This anode corrodes preferentially, liberating electrons with the same effect as the impressed current system, for example:

$$Zn \rightarrow Zn^{2+} + 2e^-$$

This system is illustrated schematically in Figure 7.3.

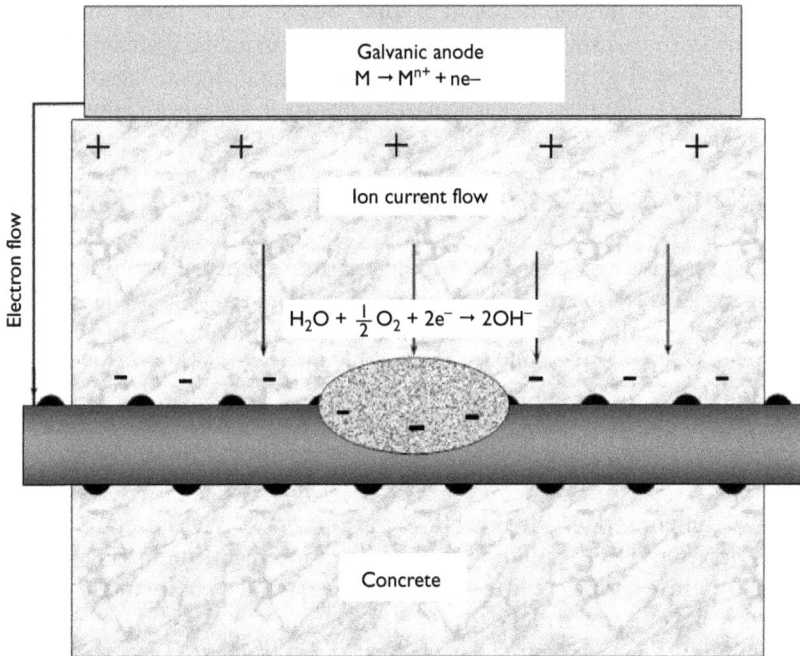

Figure 7.3 Schematic of a galvanic cathodic protection system. The steel is made cathodic by the dissolution of a suitable metal which preferentially corrodes, generating the electrons needed to sustain the cathodic reaction on the steel surface.

The same phenomenon is used in galvanizing where a coating of zinc is applied over steel to corrode preferentially, protecting the steel. However, the main restriction on this system is that the zinc has a small driving voltage when coupled to steel. This is only a few hundred millivolts, and gets smaller with actively corroding steel. While a galvanizing system puts the two metals in direct contact, with galvanic anode cathodic protection there is an electrolyte to carry the current. The resistance of the electrolyte is crucial to the performance of the system.

The resistivity of normal 'inland' concrete is high compared with sea water-exposed concrete and other aqueous, non cementitious electrolytes where zinc galvanic anodes are often used. This resistivity can be made higher by the formation of zinc oxides as it corrodes.

Unlike a free flowing electrolyte, the corrosion products are not washed away in concrete.

Aluminium and magnesium and their alloys are also used in galvanic anode cathodic protection systems for buried and submerged steel. One advantage of these alloys is that they are lighter than zinc. However, their oxides and other corrosion products are voluminous and could attack the concrete. They are therefore less attractive as anodes for embedding in concrete.

The earliest galvanic systems for reinforced concrete were installed in the United States on bridge decks. The first was in 1978 using aluminium, and from 1976 to 1980 a study was carried out on a zinc system in Illinois. Both were metal sheets under overlays on bridge decks. Despite initial negative reports, long-term performance appears to have been good with no maintenance throughout their life.

Galvanic cathodic protection systems have been used extensively since the early 1990s in Florida on prestressed concrete bridge support piles in the sea. One of the reasons the galvanic system is used there is because concrete resistivity is low due to the marine exposure conditions. The Florida systems frequently incorporate a distributed anode of zinc fixed on the atmospherically exposed concrete and bulk zinc anodes in the water which pass current through the low resistance sea water to protect the submerged area as shown in Figure 7.4.

7.3.2 Galvanic anode systems and their development

The distributed anodes used in Florida are principally electric arc sprayed zinc, a few tenths of a millimetre thick, or zinc metal mesh (Figure 7.5). Thermal sprayed zinc anodes have also been widely used in the United States for impressed current cathodic protection.

Originally, the mesh anodes were mechanically clamped to the pile. In later versions, a GRP jacket containing the zinc mesh is attached to the pile and filled with grout after connecting the zinc to the reinforcement

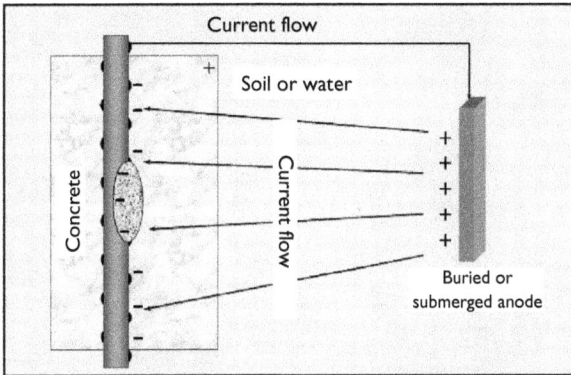

Figure 7.4 Schematic of a buried or submerged galvanic anode protecting reinforcement.

Figure 7.5 Thermal sprayed zinc being applied to a bridge substructure in the Florida Keys. Courtesy Florida DOT.

(Figure 7.6). All these systems and their performance in Florida are discussed in Kessler *et al.* (1995, 2002).

The next major development was of 'humectants' which are sprayed onto thermal sprayed zinc anode surfaces. A humectant penetrates into the concrete and increases the humidity to maintain a low electrical resistance and maintain current flow between the anode and the steel (Bennett *et al.*, 2000).

In the United Kingdom the first major development was of a proprietary anode for patch repair. This has the appearance of a 'hockey puck' consisting of a disk of zinc in a specially formulated mortar that prevents the zinc passivating. Two pairs of wires protrude from the opposite sides of the puck

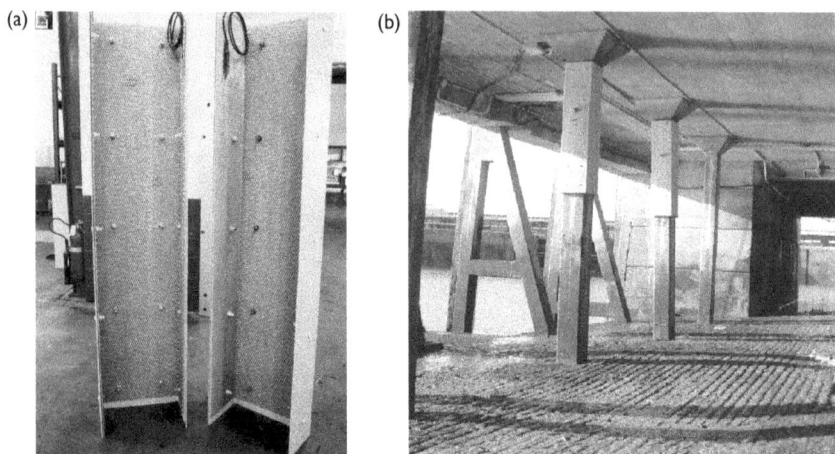

Figure 7.6 (a) A GRP jacket jetty anode assembly showing zinc mesh. Courtesy of Vector Corrosion Technologies Ltd. (b) Jacket anodes in place on a harbour in the Channel Islands, UK.

Figure 7.7 Installation of a galvanic anode in a patch repair to prevent incipient-anode-induced corrosion around the repair. Courtesy of Vector Corrosion technologies Ltd.

and are attached to the reinforcing steel exposed during the repair. By installing galvanic anodes within the repair (Figure 7.7), the 'incipient anode' effect is eliminated (see Figure 6.5).

This was then developed further into a small 'yogurt pot' sized anode with a single connecting wire that can be inserted into core holes in the

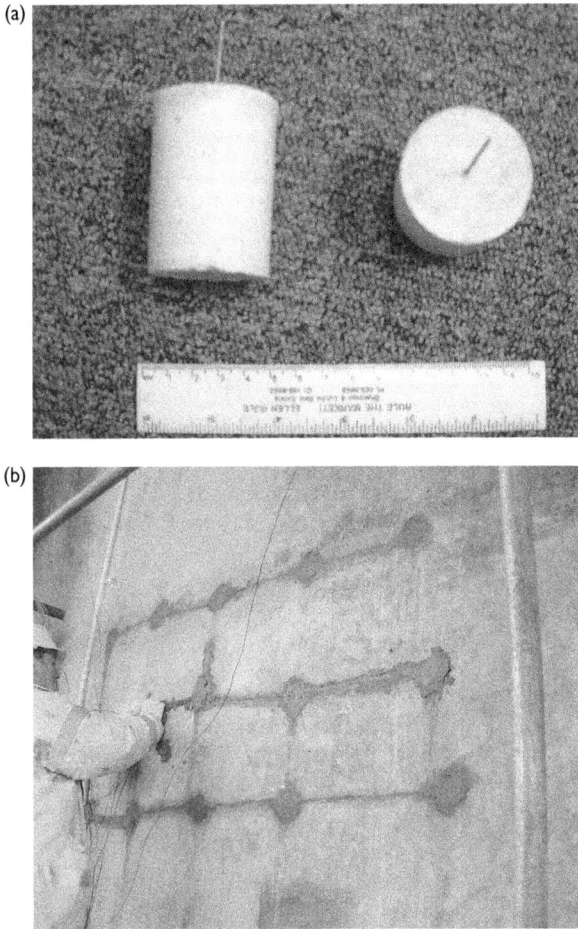

Figure 7.8 (a) Galvanic anodes for installation in cored holes. Anodes are linked with titanium wires to form a zone. (b) Galvanic anodes being installed in core holes in a wall. Courtesy Vector Corrosion Technologies Ltd.

concrete and wired together to produce a galvanic discrete anode system (Figure 7.8). They have been used predominantly on multi-storey car parks and on bridge substructures.

Since the maximum voltage that can be generated with zinc anodes is extremely unlikely to generate hydrogen embrittlement, galvanic systems have been used to protect prestressed concrete members. They are also used on fusion bonded epoxy coated steel reinforced piles as the effects of electrical discontinuity between bars is unlikely to lead to stray-current-induced corrosion as the currents and potentials are low.

Table 7.1 summarizes the different anodes and their characteristics:

- Thermal sprayed zinc – Zinc is flame or electric arc sprayed onto the concrete surface and a direct connection made to the reinforcement. It can be used as sprayed on marine splash or tidal applications. In drier locations a humectant solution of hygroscopic salts can be applied. Over 100,000 m² has been applied, in Florida alone.
- Thermal sprayed Aluminium/Zinc/Indium – A proprietary variation on thermal sprayed zinc that uses an alloy to enhance the current rather than a humectant. A 300 micrometre thick coating is applied by arc spray. Approrimately 35,000 m² had been applied up to May 2003.
- Adhesive zinc sheet – A proprietary system consisting of rolls of zinc 0.25 mm thick, 25 cm wide with a conductive gel adhesive on one side. The installed system can be painted. The edges must be sealed against water ingress as the gel can swell, leak and disbond.
- Encasement system – Initially a proprietary system developed by Florida DoT consisting of expanded zinc mesh in a permanent form, grouted onto the tidal zones of concrete piles or columns. More recently unactivated, activated and wicking anode assemblies have been used in marine and inland applications. Several thousand piles and columns have been protected with these systems.
- Probe anodes in patch repairs – Proprietary anodes approximately 65 mm diameter by 30 mm high with four wires protruding for attachment to exposed reinforcing steel in the patch. Anodes are installed on a maximum of 750 mm spacing around a patch. Approximately 200,000 were sold protecting about 100,000 m² of concrete at the end of 2002.
- Probe anodes in cored holes – Potted up zinc anode approximately 45 mm diameter by 40 or 60 mm long. Installed on 330 mm to 650 mm spacing in 50 mm diameter cored holes.

7.3.3 The merits and limitations of galvanic cathodic protection

The principle advantage of galvanic cathodic protection is its simplicity, with no power:

1   In principle this could lead to cost savings on a range of projects.
2   Its strength is also its weakness. The simplicity means there is no way of controlling it and there is no automatic flow of information on its performance in the field. However, monitoring systems can be installed if required and are required if the system is to meet the international standard BS EN ISO 12696 Cathodic protection of steel in concrete and its NACE/AMPP equivalent SP0216 Sacrificial cathodic protection of reinforcing steel in atmospherically exposed concrete structures.
3   Anodes have a limited life and need replacement at intervals that can be difficult to define. Most galvanic systems aim to provide at least ten

Table 7.1 Anode types

Anode	Environment	Application	Durability/life	Comments
Thermal sprayed zinc	Marine or anywhere (with humectant)	Requires bulky spray equipment and skilled operator	Greater than 10 year life expected. Reduces with severe exposure	Colour change to concrete is the only effect
Thermal sprayed Al/Zn/In	Marine and deicing	Requires bulky spray equipment and skilled operator	10–15 year marine exposure, 15–20 year in northern deicing	Colour change to concrete is the only effect
Adhesive zinc sheet	Not for very high wetting. Will work in very dry	No special skills apart from soldering connections	Design life is 25 to 50 years. Gel deteriorates in very wet conditions	Anode is 1–2 mm thick. Either leaves zinc metal finish or painted metal finish
Encasement jacket	Marine only, mainly columns and piles	Special skills required for grouting up jacket	Very durable. Life up to 50 year	Pile column section enlarged by 25 mm or more. Load increase. Repair can be structural
Probe anodes in cored holes	Anywhere	No special skills	15 to 25 year life. Very durable	Core holes at 330 to 650 mm centres. Requires good design
Probe anodes in patch repairs	Any patch repairs	No special skills	10 to 15 year life. Very durable	Only area around patch repairs protected

years life before the anode needs replacing. However, systems have been designed to last for 40 years or more in specific applications.

4   If the concrete is too dry the system will not work although under such conditions the risk of corrosion is likely to be reduced.

5   It has been estimated that approximately 1 million sq m of concrete have been protected with galvanic anodes between 2000 and 2008 (Broomfield and Wyatt, 2018). The market for galvanic anodes continues to grow.

6   There is now a range of anodes, which includes:

•   Thermal sprayed metals (usually zinc or alloys)
•   Encasement systems
•   Adhesive zinc sheet systems
•   Embedded anodes for extending the lives of patch repairs
•   Embedded anodes as discrete anode arrays.

7   The production of small galvanic anodes for patch repair systems is a useful adjunct to conventional patch repairs. It is not designed to provide

full cathodic protection and would require a proper cathodic protection design to ensure adequate protection and a reasonable life of the patches.

8   The driving voltage of galvanic systems is low and may be inadequate to provide full cathodic protection in very high chloride or very dry, high electrical resistance conditions.

9   The limited driving voltage may make it attractive for protecting prestressed structures liable to hydrogen embrittlement and for epoxy coated reinforcement, where electrical continuity cannot be guaranteed and the consequences of discontinuity (stray currents) are small.

7.3.4 Hybrid anodes

The most recent development in cathodic protection for steel in concrete is of hybrid anodes. The aim is to use an initial external impressed power supply to rapidly repel chlorides from the bar, repassivate it and stop any pitting attack. After that, a conventional galvanic anode maintains polarization. There are two proprietary forms of hybrid anode currently on the market. The first to be developed was a galvanic anode that could be 'charged' by an impressed current. This uses a DC power supply which can be the type of transformer rectifier used for impressed current cathodic protection or can be as simple as a car battery. Anodes are typically 18 mm in diameter and range in length from 42 to 220 mm long. The other system has a battery built into each probe anode. As the battery is depleted the galvanic anode maintains polarization. Examples are shown in Figures 7.9 and 7.10.

The advantages of the externally powered system are that it is easier to monitor performance compared to a galvanic system, and there is no permanent power supply so no ongoing costs or maintenance required. The anodes have large charge capacity with a lifetime of up to 50 years (depending on anode size). If the level of polarization falls, the impressed current can be reconnected at any time.

The advantage of the battery powered units are their simplicity, being comparable to conventional galvanic anodes with no external power supply. This means that monitoring must be installed if the system is to meet the standards on cathodic protection for steel in concrete.

7.4   The components of an impressed current cathodic protection system

The essential elements of a conventional impressed current cathodic protection system are discussed in this section and in Section 7.5.

The cathode is the reinforcing steel to be protected. It must be continuous, that is, electrically interconnected to allow current flow, and separated from the anode by an electrolyte (the pore water). If it is not electrically continuous then it must be made so by adding reinforcing bars, wiring or welding

(a)

Single-wire Connection
to the Steel

Alkali-activated Galvanic
Anode

Self-powered ICCP
System

(b)

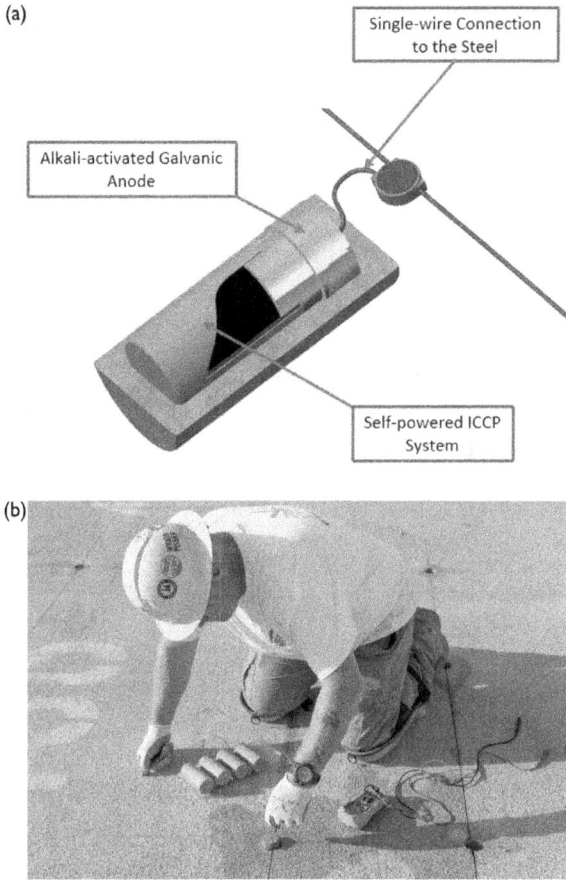

Figure 7.9 (a) Schematic of a battery powered hybrid galvanic anode. (b) Installation of battery powered hybrid anodes in a car park deck. Courtesy Vector Corrosion Technologies Ltd.

elements together. If there are 'shorts' to the anode (usually where tie wires or shallow bars come into contact with the anode placed on the concrete surface or in holes or slots in the concrete) they must be removed or the current will not flow through the steel to concrete interface where it generates the cathodic reactions (equations (7.2) and (7.3)). The electrical connection from the reinforcement to the cable leading to the power supply can be made by a number of methods including welding, brazing, tap and screw or even tie wire in the case of galvanic anodes.

Impressed current anodes have very slow or controlled consumption rates when the anodic reaction occurs on the anode surface. As the reaction consumes alkalinity (equation (7.5)) and generates acid (equation (7.6)), it

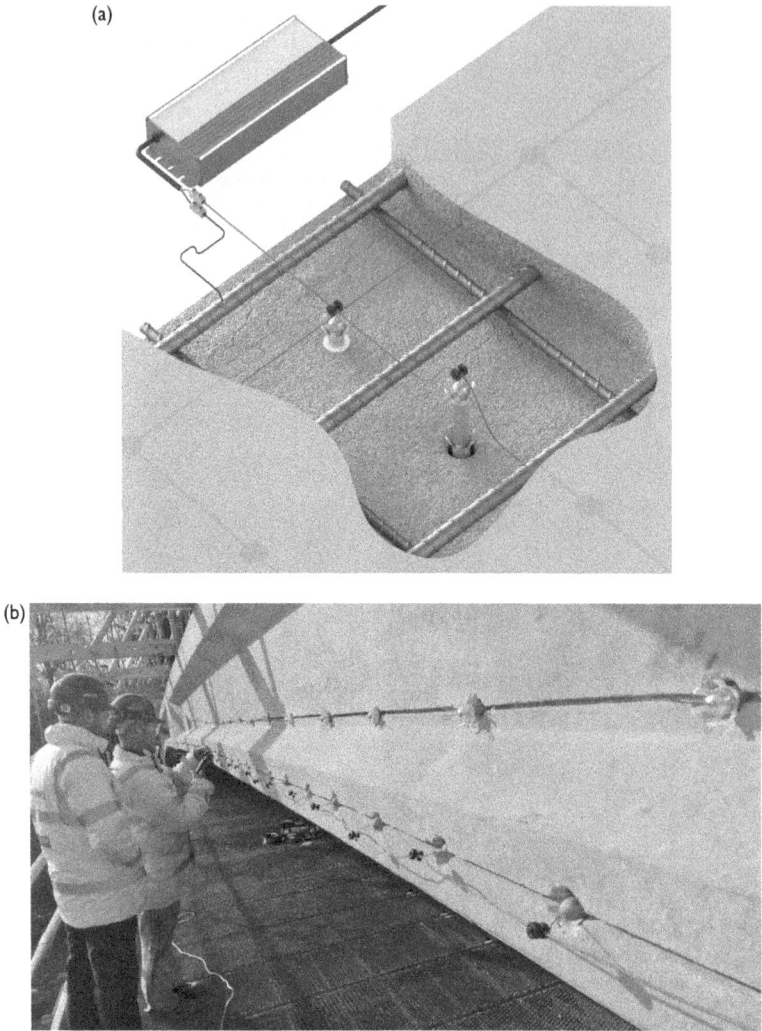

Figure 7.10 (a) Schematic of an impressed current powered hybrid galvanic anode. (b) Installation of impressed current powered hybrid anodes in a substructure. Courtesy Concrete Preservation Technologies Ltd.

can attack the anode and the concrete. The level of current is therefore important in maintaining a good anode to concrete bond. Types of anode are described later. The function of the anode is to spread the current to all areas to be protected having converted the electrical current from the transformer rectifier (DC power supply) to an ionic current that flows from the anode to the cathode so that the cathodic reaction will occur on the reinforcing steel surface, suppressing corrosion.

One important requirement is a low interfacial resistance between the anode and the concrete. Also, the electrical resistance of the anode system should be proportionately lower than the combination of the interfacial resistance, the concrete cover resistance, and the steel to concrete resistance, otherwise the current will not distribute evenly through the concrete to the steel. For atmospherically exposed reinforced concrete structures the anode is usually a 'distributed anode system', such as a conductive coating on the surface, an expanded metal mesh across the surface encased in a concrete overlay, strips of anode in slots across the surface or a series of small 'point' or 'discrete' anodes embedded in the concrete cover or among the rebars.

As gases are evolved at the anode (equations (7.4) and (7.5)) the distributed anodes such as coatings or encased anodes must be gas permeable.

The electrolyte is the medium through which the ionic current flows from the anode to the cathode. It can be soil or sea water for pipelines or ships. In the case of atmospherically exposed reinforced concrete structures it is the concrete pore water. In the case of buried or submerged concrete it will be firstly soil or water and then the concrete pore water as shown in Figure 7.4 for a galvanic system. Our discussion will concentrate on atmospherically exposed concrete structures.

The tortuosity and small size of the pores in the concrete gives a higher resistivity on concrete structures than in marine or soil CP systems. Added to this is the fact that not all the pores are 100% full, or even lined with water. This makes a very unusual highly oxygenated, stagnant, high resistivity, alkaline medium. This contributes to the unique requirements of cathodic protection systems for atmospherically exposed reinforced concrete structures.

Problems can arise if the concrete around the anode dries out. Initially this gives rise to a requirement for a higher driving voltage (10–15 V instead of the usual 1–5 V). It is usually assumed that if the concrete dries out much more than this, then the concrete around the steel is too dry for corrosion to occur. This may not be true, but will not matter if the drying out is short term (say less than a month or so during a hot dry summer). Problems have been observed on indoor parking garages heated in the winter. Salty ice melts on the deck, the salt transports into the concrete in the water and accumulates around the steel but the moisture around the anode dries out. The anode encounters a high resistance and cannot pass sufficient current to control the corrosion at the moist, salty reinforcement surface, corrosion ensues.

The Transformer/Rectifier (T/R or rectifier) is the DC power supply that transforms mains AC to a lower voltage and rectifies it to DC. The positive terminal is connected to the anode and the negative to the cathode. The level of the output is controlled as described later. T/R's can be run at constant voltage, constant current or constant potential (against a half cell). They can be adjusted manually or automatically by circuitry or computer control or remotely using a telephone line and modem link or a similar remote connection as described later.

Transformer/rectifiers for conventional cathodic protection systems can be very large and powerful, capable of delivering hundreds of amps, with oil cooled transformers. However, for steel in concrete the requirements are more modest. A graph of current requirement vs. chloride content (Figure 7.15) is discussed under 'Control Criteria', Section 7.5. Most systems are designed for a current density of about 10–20 mA per square metre of steel surface for actively corroding structures and for 0.2–2.0 mA/m^2 for new structures where there is no pitting and so no need to passivate pits (see Section 7.2). There may be a requirement for allowances to be made for the current flow to lower layers of steel. Allowances must also be made for the voltage drops down the connecting cables.

The problem with selecting a transformer rectifier is ensuring that it is powerful enough, but not so powerful that all adjustments are made in the first 10% or less of the output. This can lead to difficulties in control and to inefficiency in output, with electrical energy lost as heat. The electrical engineering of transformer rectifiers is covered in other texts on cathodic protection, see for instance Chess and Broomfield (2014), and will not be discussed in detail here.

In general terms, the systems for protection of steel in concrete are generally full wave rectifiers with smoothing to minimize interference and any possible adverse effects on the anode. In practice, for steel in concrete with a low DC output, switch mode supplies are generally preferred. The different types of power supplies are discussed in Chess and Broomfield (2014). A continuously variable output is usually specified. Most cathodic protection systems are run under constant current control, although constant voltage (or an option for both methods) is sometimes specified. Control by constant half cell potential against an embedded reference electrode is rarely specified for steel in atmospherically exposed concrete but may be applied to buried or submerged parts of structures.

Transformer rectifiers must be adequately protected from the elements. If they are sited outside they are usually in steel enclosures (suitably corrosion protected!) with adequate ventilation and protection against vandalism or other damage. They often incorporate lightning arrestors and other protection for the public and the inspector of the system, with suitable fusing and earth protection.

Other sources of DC power are available in principle but are usually too expensive for use in routine installations. Wind or water turbines, solar arrays and even batteries have been proposed and experimented with, but few have been used in large scale practical applications.

Reference electrodes or other small probes may be embedded in the concrete. These measure the effect of the CP current and enable operators to set correct current and voltage levels for the system. These are specifically designed for long-term durability in an alkaline concrete environment. The commonly used designs are double junction electrodes to minimize long-term contamination or leakage. They are usually silver/silver chloride potassium chloride or a proprietary manganese/manganese dioxide type.

7.4.1 Selection of anodes and early anode development

The anode is a critical part of the cathodic protection system. It is usually the most expensive item and has the highest cost of installation with major disruption to the operation of the structure. It is crucial that the correct anode is chosen and that it is applied properly.

The first anodes, developed by Stratfull, were silicon iron primary anodes in contact with a conductive coke breeze asphalt overlay (the secondary anode). This design was based upon CP designs for pipelines where a silicon iron anode is embedded in a carbon coke breeze backfill to give a large contact area and low resistance. The anode is then linked, via the transformer rectifier, to the pipeline to be protected. A modified form of the conductive asphalt system was used for several decades in Ontario, Canada. In some cases carbon anodes have been used instead of silicon iron. The next development was a conductive polymer grout (carbon loaded resin) put into parallel slots cut into the surface. These two systems were only suitable for decks and have rarely been applied outside North America. The aim of the slot system was to reduce the weight of the anode and to eliminate the change in deck height. However, this system requires good concrete cover to the reinforcing steel to avoid short circuits between the anode and the reinforcing steel.

One of the first commercial, proprietary anodes for decks was a flexible cable with a conductive plastic round a copper conductor. The cable was 'woven' across the deck and then a concrete overlay applied. It was also used on substructures with a sprayed concrete overlay. This system was very popular in the 1980s but unfortunately the conductive plastic started to fail after about five years of service. This led to attack of the copper and, in some cases, expansion of the plastic which delaminated and spalled the concrete overlay. This system is no longer available for CP of steel in concrete.

7.4.2 Conductive organic carbon loaded paints

Conductive organic coatings consist of a conductive carbon 'pigment' in a suitable binder. The most widely used systems have been solvent based. Water-based systems are now widely used. One proprietary system uses a nickel coated carbon fibre rather than flaked graphite. In addition, coatings with conductive, inorganic binders are now available.

Coating anodes were developed from the carbon loaded paints used in electronics and specialist electrical applications and developed for space heating in the 1960s. Poor results for the application of these anode systems were reported in North America in the 1980s, with the first generation of coatings debonding and flaking off.

Research carried out in the United Kingdom for the DoT led to development of a system that was durable in UK conditions. The North American manufacturers developed their materials further and improvements in concrete surface preparation have lead to more durable systems. A range of coatings is now available.

Coatings do not have the durability of the coated titanium-based anode systems. However, they are less expensive, easier to apply and can be repaired and maintained easily. They are not suitable for continuously wetted abraded or trafficked surfaces. They require excellent surface preparation to get good adhesion (see Came, 2018).

All anode systems require an electrical connection. Given their modest conductivity this usually consists of a series of wires (platinized titanium, platinized niobium copper, carbon fibres or other comparatively inert materials) running through the coating at a separation of 0.5–1.0 m. Unfortunately, the nomenclature used for the early anode systems has been misapplied to coatings. In the conductive asphalt system and the pipeline system it is based upon, the silicon iron (or carbon) anode is called the primary anode. The coke breeze material (backfill or asphalt) is called the secondary anode. Likewise the connection wire in the paint system is often referred to as the 'primary anode' and the coating as the 'secondary anode'. This considerably inflates the significance of the connection wire and diminishes the importance of the coating which is really the anode. Examples of conductive coating anodes are shown in Figure 7.11.

7.4.2.1 Installability

Installation is by airless spray to accessible areas and by roller or brush to inaccessible locations such as between the pairs of shear walls. Surface is prepared by a light grit blast or similar, to achieve an adherent surface. See Came (2018). Requirements for a dry concrete surface 3°C or higher means that winter application can be difficult as even with enclosures and space heating, the cold concrete surface can create condensation.

7.4.2.2 Suitability for particular locations

The coatings are most suitable for application to dry or near dry elements or faces. They do not tolerate water during installation or continuous wetting during operation, they are easily applied and maintained but have a limited life.

7.4.2.3 Life and cost

According to anode suppliers, as reported in Broomfield (2018) these anodes have a life of up to 15 years and a cost of £100 to 300/m² for UK

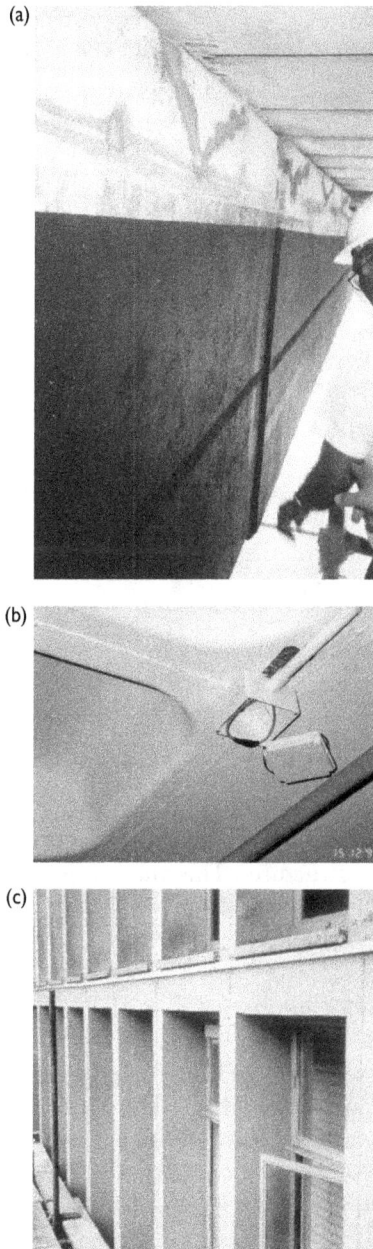

Figure 7.11 (a) A conductive paint anode applied to a cross head on the M4 elevated section in London. A solvent based and a water-based coating were trialled. Note the step out where cover was increased to prevent short circuits between the skewed reinforcement cage and the conductive coating. Probe anodes are installed under the bearing shelf. Individual connections were made to the dowel bars into the longitudinal beams to ensure electrical continuity and protection. (b) A very early application of a conductive coating anode being applied to a building in 1986. Courtesy Vinci Technology Centre. (c) After application of the cosmetic top coat. Courtesy Vinci Technology Centre.

highway structures. However, on non-highway structures costs may be significantly lower due to easier access and other cost issues of working on live highway structures.

7.4.2.4   Impact on structure

With a decorative/protective coating the anode has minimal impact on the structure in terms of loading, change of profile or appearance. However, as the decorative coating deteriorates, the black conductive coating shows through.

7.4.2.5   Performance

Originally specified with a 10- to 15-year performance, the oldest systems on the Midland Links were replaced after about 20 years. Some areas had performed badly due to excessive water exposure. Some poor quality products had been used and in some cases inadequate surface preparation has given problems. However, on the whole these anodes significantly exceeded the 15-year life expectancy.

7.4.2.6   Extent of use

This was the most widely used anode in the United Kingdom for many years with over 600 cross heads protected on the Midland Links motorways in Birmingham since the first trials in 1986–1987. They are also widely used on building and other structures from water towers to multi-storey car parking structures. It was also widely used in the United States and Canada, particularly on parking structures. This anode is less widely used now in UK and North America but is supplied and applied by CP companies based in mainland Europe.

7.4.2.7   Sources of further information

The US National Association of Corrosion Engineers (NACE), recently renamed the Association for Materials Protection and Performance (AMPP), has published a test method for evaluating coatings. TM01105-2018 *Evaluation of coatings containing conductive carbon additives for use as an anode on atmospherically exposed reinforced concrete.*

7.4.3   Thermal sprayed zinc

Given the early problems with conductive coatings in North America, Caltrans looked for an alternative. They experimented with electric arc and flame sprayed metals and found zinc to be most effective.

Many arc sprayed zinc systems have been applied. They are particularly popular in marine substructure conditions where it is difficult to apply other anodes. Florida Department of Transportation has applied several such systems and Oregon DOT have sprayed several very large bridge substructures over the past few decades. The Yaquina Bay bridge has a $26,500 \, m^{-2}$ system applied and Cape Creek is about $13,000 \, m^{-2}$. The electric arc sprayed system is preferred over arc spraying in North America due to its rapid deposition rates. Figure 7.5 shows the arc spraying process for a galvanic anode system in the Florida Keys.

One of the few impressed current zinc systems in the United Kingdom that the author knows of and was involved with, is shown in Figure 7.10. There has been some concern about the rise in resistance seen on some systems. This may be due to a build up of corrosion products between the zinc and the concrete, or to treatment of the zinc after application to protect it from atmospheric corrosion. Zinc, of course, is not inert and is consumed by corrosion from the atmosphere and water impingement. The anodic reaction also consumes the zinc and gives rise to the formation of oxides and sulphates at the anode/concrete interface which may increase the electrical resistance between anode and cathode.

The use of high temperature spraying gives a very porous, open coating made up of small droplets of zinc metal. A properly applied coating has excellent adhesion to the concrete. The zinc coating looks similar to concrete and there is no need for extra protective or cosmetic coatings. The system is well established in the United States both as an impressed current and as a galvanic anode (see Figure 7.5).

Among the highway agencies in North America there has been a preference for arc sprayed zinc systems for protecting bridge substructures. In Europe the preference has been for conductive paints which have been applied extensively to multi-storey parking structures as well as some buildings and bridge substructures. One reason they were successful in the United Kingdom on highway structures is that guttering was applied below leaking joints to minimize the amount of water impinging directly on the coatings. This has extended their life. In North America this has not been done in many cases so the carbon based coatings do not last as well as the zinc. Also the zinc can be applied to damp concrete surfaces, making it suitable for marine applications, and where the water continues to impinge on the concrete surface.

As zinc coatings are highly conductive electrical connection is made via a metal plate fixed to the surface and the zinc is sprayed over it. Very few connections are needed, but at least two should be applied in each zone.

7.4.3.1 Description of the anode

Pure zinc (99.99%) is sprayed using flame or electric arc. It is widely used in the United States where conductive organic coatings were not found to be

Figure 7.12 Thermal sprayed zinc applied to the leaf piers of Golden Fleece Interchange on the M6, UK. Probe anodes are applied to the cantilevered ends, the bearing shelf and the diaphragms between the longitudinal steel.

successful on highway substructures (this is probably due to a combination of poor surface preparation, lack of control of water leakage onto the coating and more extreme thermal cycling). The zinc coating is far thinner than a conductive paint. The concrete is left with a slight metallic grey sheen (see Figure 7.12).

7.4.3.2   Variations on the basic description

The use of such anodes in galvanic mode is widespread on Florida coastal bridges. A high solids spray applied zinc coating has been offered in Europe but there is little independent published experience of its application or performance. Elsewhere in the corrosion industry it has been found that solvent- or water-based zinc coatings with a high enough zinc content to be conductive have poor durability.

7.4.3.3   Installability

The material is non-proprietary. However, it requires specialist applicators and spray equipment for external application (most metal spraying is done in factories and workshops). Some is done on site to repair galvanizing. Few UK applicators have experience of applying it to concrete. The spray is more forgiving than organic coatings. As the zinc is a molten spray it can tolerate colder conditions and damp concrete. As the coating is highly conductive it cannot tolerate short circuits to the steel caused by tramp steel or tie wires. However, a simple electrical circuit with an audible alarm can be set up to warn of shorts between the reinforcement cage and the anode as it is applied. It can still be difficult to precisely locate the short.

Zinc vapour can lead to 'zinc flu' if breathed in. Therefore, all operatives must be suitably protected and the area shrouded to contain the zinc.

7.4.3.4 Suitability for particular locations

The coating is more tolerant of water after application than organic or inorganic binder coatings. However, it requires bulky equipment and cannot be applied in any other way. There are also the health and safety issues to be considered.

7.4.3.5 Life and cost

According to Broomfield (2018) these anodes have a life of up to 25 years and a cost of £200–£400/m^2 for UK highway bridges. However, costs may be far lower for non highway structures.

7.4.3.6 Impact on structure

Like a coating there is little visual or other impact. The grey colour means it has limited acceptability on buildings, however, compatible cosmetic top coats are available.

7.4.3.7 Performance

According to Broomfield (2018) a typical life estimate is up to 25 years. There is a limit to the thickness of zinc that can be applied to concrete which is part of the limit on the life before re-application is required.

7.4.3.8 Extent of use

Several hundreds of thousands of square metres have been applied to bridges in the USA. The largest installations have been in Oregon where a number of historic landmark bridges were protected with this anode applied to the substructure. There have been other large installations on inland bridges elsewhere in the USA and Canada. However, its use has declined in recent years compared to titanium based anodes. This is partly because of the use of specialized equipment for spraying, the expertise needed for spraying concrete as opposed to steel, where it is more widely used, and the health and safety issues of spraying zinc.

7.4.3.9   Sources of further guidance information

American Welding Society. Specification for thermal spraying zinc anodes on reinforced concrete. AWS/ANSI Standard. 2016; AWS C2.20/C2.20M:2016.

7.4.4   Coated titanium expanded mesh in a concrete overlay

One of the most successful commercial anodes is the expanded titanium mesh with an activated precious or mixed metal oxide coating. This also comes in the form of an expanded titanium mesh, strips and other configurations. It is fixed onto the surface, usually with plastic fixings and a cementitious overlay applied.

For soffit or vertical surfaces sprayed concrete is normally used to overlay the anode. As stated in the section on patch repairs, sprayed concrete application requires considerable expertise to get a good bond between the overlay and the parent concrete. This is done by preparing the original surface (grit blasting scabbling or similar roughening and cleaning) and standing the mesh off the surface so that concrete rebound from the anode is minimized. The highest standards of 'on site' quality assurance and control are needed to ensure 100% adhesion of the overlay.

The mesh has also been mounted in permanent form work and then grouted up by pumping in concrete or mortar from above or below. This avoids the quality control problems of sprayed concrete but gives more engineering problems and can increase the cost of the permanent form work.

Electrical connections are made to the anode with strips of titanium welded to the mesh.

7.4.4.1   Description of the anode

This anode consists of an expanded titanium mesh with a catalytic mixed metal oxide coating. The anode is fixed to the concrete surface and then overlaid with concrete. This is usually dry spray shotcrete on vertical and soffit surfaces or cast on top surfaces and decks. It is also available in a number of mesh sizes and as a ribbon (described separately).

7.4.4.2   Installability

The main issue is the bond between overlay and parent concrete (see Figure 7.13b). High QA/QC procedures and an experienced nozzle man are required. Delaminations have occurred within the overlay but do not appear to have a major impact on the performance of the system or its durability. The application is forgiving of moisture during application but must avoid freezing temperatures during application and curing.

7.4.4.3   Suitability for particular locations

This anode system is one of the most durable and can be applied in any condition from underground and underwater to dry atmospheric exposure where the change in appearance, profile and dead load is acceptable. It has been widely used on trafficked decks and on substructures, particularly in marine exposure conditions.

7.4.4.4   Life and cost

According to Broomfield (2018) these anodes have a life of up to 120 years and a cost of £200 to 400/m² for UK highway bridges. However, costs can be significantly lower for non highway structures not suffering from the access issues of live highway structures.

7.4.4.5   Impact on structure

The overlay must be about 25 mm thick so it changes the profile, loading and clearances as well as the appearance. Clearance may be an issue round bearings. It is possible to get a reasonable appearance with sharp corners, etc. by 'flashing' the surface.

7.4.4.6   Performance

This anode is used for long life, low maintenance applications especially in aggressive environments. The main problems are with the overlay where adhesion can be a problem (see Figure 7.13b)

7.4.4.7   Extent of use, etc.

This is probably the most widely used and robust of anodes when properly applied. It accounts for about 50% of the several million square metres of impressed current cathodic protection applied worldwide (Broomfield and Wyatt, 2018). It is widely used on bridge decks, marine substructures and in any aggressive environment.

7.4.4.8   Sources of further information

There is a NACE/AMPP Standard test method for testing embeddable impressed current anodes. TM0294. *Testing of embeddable impressed current anodes for use in cathodic protection of atmospherically exposed steel-reinforced concrete.* There is also an equivalent ISO standard based on the NACE test method (ISO, 2018). This was developed to assess the performance of the coated titanium anode equivalent to a 40-year life. It does not test the overlay.

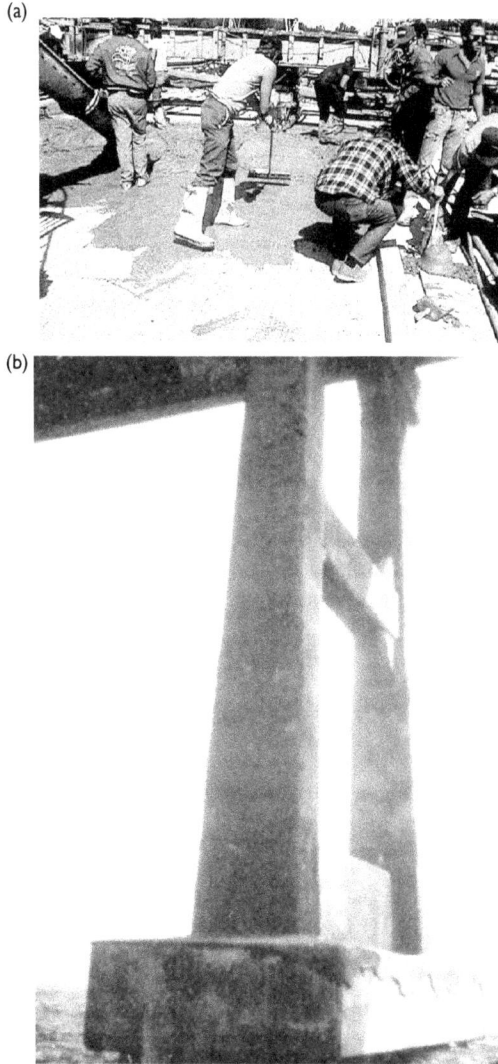

Figure 7.13 (a) A mixed metal oxide coated titanium mesh anode being overlaid on a bridge in Baltimore MD. Courtesy of De Nora Group. (b) Extreme example of an overlay failure on a titanium mesh anode. Oregon Inlet N. Carolina USA. The whole bridge was swept away in a storm a few years after the author took this photograph.

7.4.5   Coated titanium expanded mesh ribbon mortared into slots chased into the concrete

7.4.5.1   Description of the anode

This anode consists of an expanded titanium mesh with a catalytic mixed metal oxide coating. The anode is in the form of a ribbon about 13 mm wide. Slots are cut into the concrete typically on 300 mm centres. These are filled with mortar and the ribbon pushed into the slots as shown in Figure 7.14(a).

An alternative system for mounting the anodes is a proprietary 'cassette' system with a ribbon encased in a glass reinforced plastic form with a contact felt and proprietary hygroscopic fluid. The assembly is fixed to the concrete surface with stainless steel screws. This was designed for rapid installation above mid tide level on marine exposed structures and has also been used on inland bridges. See Figure 7.14b. (Atkins, 2010; Bruekner *et al.*, 2012)

7.4.5.2   Installability

The main requirement for the ribbon installed in the concrete cover is for adequate cover to avoid short circuits or excessive current drain. Slots can run round obstacles. The application is forgiving of moisture during application but must avoid freezing temperatures during application and curing. The cassette system is designed for simplicity and ease of mounting with screws and wall plugs. As long as the surface areas where the anode is installed are free of contaminants and of tie wire or other metallic items that might cause short circuits it will operate effectively.

7.4.5.3   Suitability for particular locations

This anode system is one of the most durable and can be applied in any condition from underground and underwater to dry atmospheric exposure. It is also widely used on historic buildings with steel frames behind brick or stone façades where the ribbon can be inserted in the mortar joints as shown in Figure 7.14(c). The cassette is designed for use above water but has been used inland (Atkins, 2010).

7.4.5.4   Life and cost

According to Broomfield (2018), the ribbon anodes installed into slots in the concrete have a life of up to 120 years and a cost of £150 to 400/m² for UK highway bridges. However costs may be significantly lower for non-highway structures.

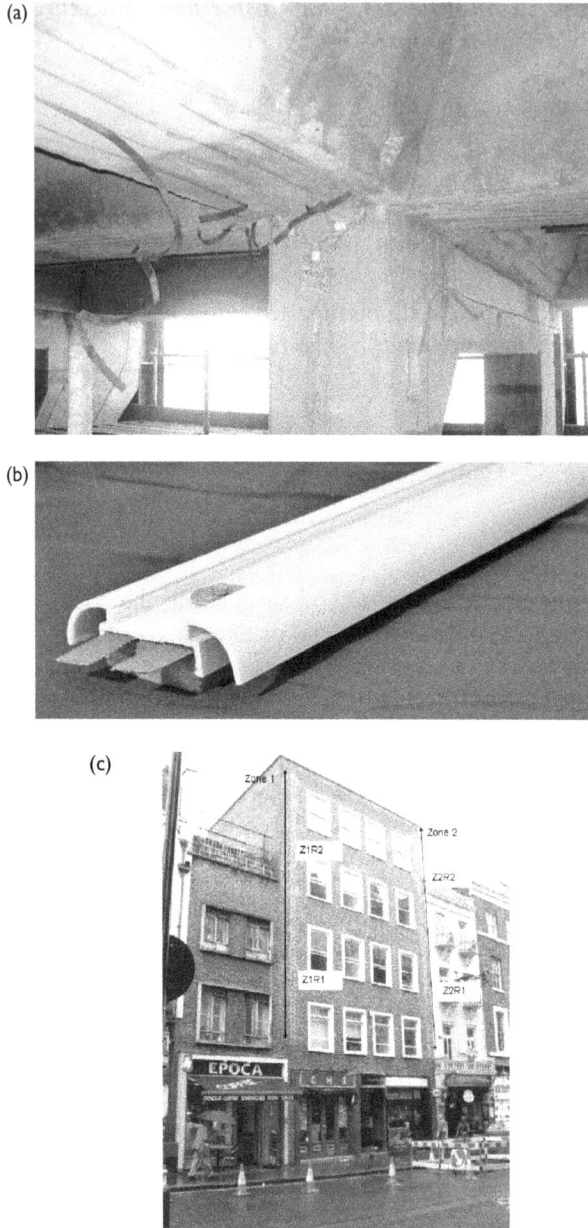

Figure 7.14 (a) Ribbon anodes being applied to the underside of a jetty in Jersey. Certain areas of the soffit, beams and all of the columns plus support trestles were protected using ribbon anodes. The tide rises to near soffit level and there is an 11 m tidal range. (b) Cassette anode with mixed metal oxide coated titanium ribbon in a glass reinforced concrete form with felt contact soaked in hygroscopic fluid. (c) 55–57 St Martin's Lane London showing the front two zones on a brick building with steel columns after installation of ribbon anode system the reference electrode locations are indicated by Z1R1, etc.

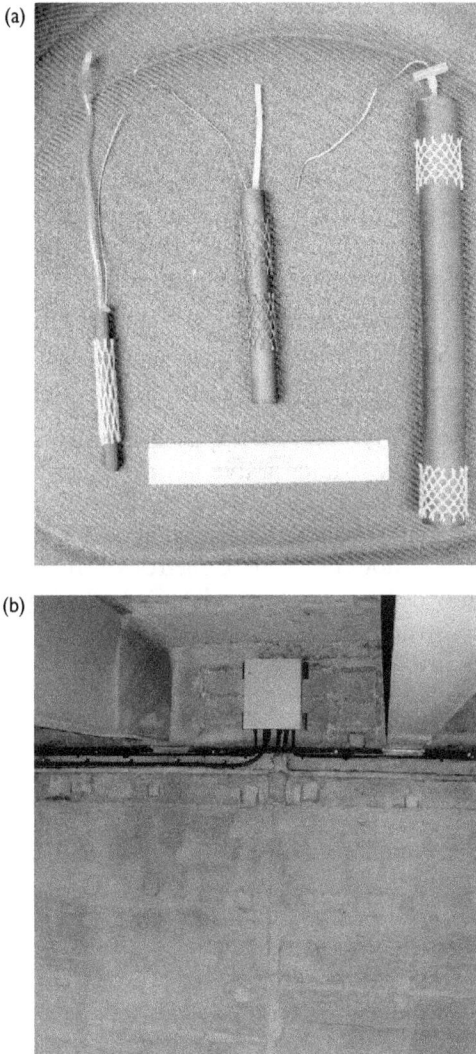

Figure 7.15 (a) Ebonex conductive ceramic probe anodes showing gas vent tubes. (b) Probe anodes on Golden Fleece Interchange. In diaphragm and below bearing shelf. Square holes are shutter bolts holes removed. Zinc anode below. See Figure 7.11(a) for probe anodes on M4 and Figure 7.12 for general view of Golden Fleece.

7.4.5.5   Impact on structure

Inevitably there is a 'stripy' appearance after being embedded in the cover of concrete structures, but no change to loadings, profiles or clearances. On brick or stone building façades the new mortar may show depending on the extent of repointing (Figure 7.14(c)). The cassette system has minimal added load but is potentially vulnerable to impact damage. There will be visual impact of the appearance of the cassettes.

7.4.5.6 Performance

This anode has performed well in long life, low maintenance applications especially in aggressive environments and steel framed masonry and brick clad buildings and monuments. However, there has been evidence of localized acid attack where ribbons are not embedded deep enough in the cover concrete and where used in tidal areas.

7.4.5.7 Extent of use

The anode is used where increase in dead load, change in profile or application of an overlay is undesirable but the durability of the titanium mesh anode is required. It requires good cover to the steel and can be more expensive to install than the shotcreted mesh anode. It is not as widely used but there are several thousand square metres of application around the world. It has been widely used in 'cathodic prevention' systems installed on new or nearly new structures.

Ribbon anodes are also increasingly popular on historic steel framed masonry or brick clad structures where they can be fitted in the mortar joints (see Figure 7.14(b)).

7.4.5.8 Sources of further information

The NACE Standard test method for testing embeddable impressed current anodes applies to this anode. TM0294. *Testing of embeddable impressed current anodes for use in cathodic protection of atmospherically exposed steel-reinforced concrete.* ISO (2018) can also be used to certify this anode. The NACE test method and the ISO standard which was based on the pre-existing NACE Test method were developed to assess the performance of the coated titanium equivalent to a 40-year life or more. They do not test the grout that the ribbon is embedded in or the cassette system.

7.4.6 Coated titanium or conductive ceramic rods, tubes, etc. in holes drilled into the concrete

7.4.6.1 Description of the anode

This anode consists of rods or tubes of titanium with a catalytic mixed metal oxide coating or conductive ceramic. Anodes are connected by titanium wires in strings to form zones. Some anodes include an optional gas vent for installation below waterproofing layers or in water saturated conditions where gas pressure might build up a 'blow off' pieces of concrete.

Different sizes and outputs are available. Sizes include a cylinder of diameter ranging from 7 to 28 mm and length from 100 to 600 mm (Figure 7.15(a)). Current outputs range from 2.0 to 50 mA. One proprietary system incorporates a balance resistor in the anode head minimizing the risk of 'current dumping'. Other anodes of a simple cylinder of mixed metal oxide coated titanium, rolled up mesh or short lengths of ribbon have also been used.

7.4.6.2 Installability

In principle installation is simple. Holes are drilled into the concrete and chases cut between holes. The holes and chases are filled with cementitious grout. The anodes are inserted into the grout and wires run between anodes in the chases. In practice there is a risk of hitting steel and therefore having to re-drill some holes, particularly with congested steel.

7.4.6.3 Suitability for particular locations

Probe anodes have been most widely used on highway structures to provide cathodic protection current to steel that cannot be reached from a surface anode. They have been applied to half joints and to provide current to two faces of shear walls, etc. where the inside face is inaccessible (Figures 7.15(b), 7.12). They are also increasingly applied on historic steel framed masonry or brick clad structures where they can be fitted in the mortar joints. Another application has been to 'stitch' them either side of cracks where salt water has leaked though causing localized chloride contamination.

7.4.6.4 Life and cost

Broomfield (2018) says these anodes have a life of up to 50 years and a cost of £160 to £400/m². Life can be extended by closer spacing of anodes running at lower currents.

7.4.6.5 Impact on structure

The drilled holes and chases have an impact on the appearance but none on the profile or clearances. The act of drilling holes could require a structural evaluation although spacing of 300–500 mm minimizes structural impact in most cases. Placing anodes in mortar joins of masonry or brick clad steel framed buildings gives minimal visual impact.

7.4.6.6 Performance

Has performed well in bridge substructures and steel framed masonry and brick clad buildings and monuments.

7.4.6.7 Extent of use

Anodes not much used in United States but have found extensive use in Europe on buildings, swimming pools, bridges and other structures. Very wide use in historic masonry or brick clad steel framed buildings and monuments.

7.4.6.8 Sources of further information

The NACE Standard test method for testing embeddable impressed current anodes applies to this anode. TM0294-2016. *Testing of embeddable impressed current anodes for use in cathodic protection of atmospherically exposed steel-reinforced concrete.* This was developed to assess the performance of the coated titanium equivalent to a 40-year life. It does not test the grout that the anode is embedded in. ISO (2018) uses the same test methodology as TM 0294.

7.4.7 Conductive cementitious overlay containing nickel plated carbon fibres

7.4.7.1 Description of the anode

This is a wet sprayed mortar containing nickel plated carbon fibres to achieve conductivity (Figure 7.16). It requires a primary anode, usually the titanium mesh ribbon. This is a proprietary anode and there are no variations.

7.4.7.2 Installability

The usual caveats apply concerning any sprayed cementitious overlay. However, if the correct spray equipment is used and properly set up, the anode has excellent adhesion properties and delaminations are very rare. The main problem is getting an acceptable finish.

7.4.7.3 Suitability for particular locations

The anode is durable and has excellent adhesion but gives a poor surface finish so may require a render or overlay for cosmetic purposes. Suitable for hidden areas such as wharf substructures.

7.4.7.4 Life and cost

According to Broomfield (2018) these anodes have a life of up to 25 years and a cost of £150 to 250/m^2 on highway structures. Costs may be lower on non highway structures due to fewer constraints due to working with moving traffic etc.

Figure 7.16 Conductive mortar anode applied to a marine bridge substructure.

7.4.7.5   Impact on structure

The overlay is typically 12 mm thick. This has an impact on the profile and clearances. The fibre concrete leaves a poor finish. However, the experience and expertise of the applicator can mitigate this. The rough finished surface can accumulate debris. It is also dark grey in colour. A rendered finish is sometimes applied. This can lead to a very attractive finish but further build up of the profile and dead load as well as further reduction on clearances.

7.4.7.6   Performance

Excellent adhesion of anode to concrete.

7.5   Cathodic protection system design

In the following discussion we do not intend to show the reader how to design a cathodic protection system from scratch. We will show some of the major processes that a qualified and experienced corrosion expert will go through in designing a system, showing how decisions are arrived at concerning the components and the system performance. Design should always be undertaken by a suitably qualified and experienced expert. The new revision of BS EN ISO 12696 has an informative (non-mandatory) Annexe B on the design requirements of both impressed current and galvanic systems to conform to the standard.

7.5.1 Choosing your anode

If CP is the chosen rehabilitation methodology then the correct choice of anode is vital. For applications where the life is less then 20 years and suitable anodes are available, galvanic cathodic protection may be preferred. For longer lives, impressed current systems are more likely to be suitable assuming that power is available and maintenance will be conducted. When it comes to individual anode choice then the table in Broomfield (2018) summarizes the merits and limitations discussed in Sections 7.3.1 to 7.3.7.

If the structure is a wearing surface then coatings are usually excluded. This usually leads to the use of one of the titanium configurations in an overlay or titanium ribbons in slots.

The main restrictions on the overlay type systems are because of the cementitious overlay:

- they increase the dead load on the structure;
- they change the profile and can reduce clearances;
- the overlay can be difficult to apply in restricted areas or with complicated geometries, especially when applied as a sprayed concrete overlay.

For mainly dry substructures, buildings and other soffit or vertical applications coatings may be suitable. These have a lower life than the titanium-based systems but are easily maintained and are cheaper. They are often more cosmetically attractive for buildings as masonry paint type overcoats are supplied. The thermal sprayed zinc system can be used instead of the paint type coating. As the metal is sprayed hot it can dry out a damp surface enabling it to bond, unlike the paint type coatings. It is therefore preferred in marine or regularly wetted environments. The newer water-based organic coatings are slightly more moisture tolerant than the solvent-based coatings.

The advantages of the coatings are:

- negligible increase in dead load;
- can be applied to any geometry;
- choice of decorative finishes;
- cheap and simple to repair or replace organic coatings.

Its restrictions are:

- limited resistance to wear and moisture;
- limited lifetime (about 15 years in most applications);
- requires rigorous health and safety control (thermal sprayed zinc). In a report on the service life of over 100 impressed current CP systems in the Netherlands Polder *et al.* (2014) found that conductive coating systems had a lower life cycle cost up until about 20 years and then a MMO Ti system became more cost effective.

Galvanic anodes are extensively used in continuously wetted environments. They need less maintenance than the impressed current systems. One problem with the zinc system is the environmental impact of the during spraying. If enclosure is required during spraying, the costs can be very high.

The conductive ceramic and other tube or rod anodes can be used to apply current locally to inaccessible areas in conjunction with other anodes or they can be drilled into a series of holes running up a column or similar structure. Their main problems are ensuring that the reinforcing steel does not shield the current, making sure that the anodes do not contact the reinforcing steel or overprotect it locally, and ensuring that the wiring to series of anodes is not too complicated. The mesh and ribbon anodes are preferred where maximum life and durability are required.

7.5.2 Transformer/rectifiers and control systems

This is another vital part of an impressed current system. The T/R must be rugged and reliable with minimal maintenance requirements. It should be easy to maintain with good instruction manuals, circuit diagrams for maintenance and easy access to fuses and other consumable and replaceable components. Compared with pipeline or marine CP applications (steel piles, etc.) the power demand is modest. Most steel in concrete needs less than 10 mA·m^{-2} to provide protection, usually at less than 10 V. Typically, the power for a 100 watt light bulb will protect 10,000 m^2. This means that a single phase, air cooled T/R will usually protect even the largest structure and power consumption is rarely an economic concern.

The required output of the transformer/rectifier and of the anode can be estimated from Figure 7.17 taken from Bennett and Turk (1994). Although the work in the report was done to try to develop a simple constant current criterion for cathodic protection, it has never been validated outside the laboratory. It should also be remembered that the areas of highest corrosion and highest chloride will have been repaired so the current demand will be reduced in proportion to the area repaired. A case study of the calculation of current densities and the resultant T/R outputs was published by

Broomfield (2004). The subject was also looked at more rigorously and mathematically by Hussainan *et al.* (2002).

Most monitoring and control systems for cathodic protection of steel in concrete are automated, collecting data and storing it in a computer controlled system. An additional option is to include remote control and monitoring. This is increasingly done via an internet connection and a website that connects the operator to the data and allows adjustments to the system. In principle this is more future proof than a direct connection to the operator's computer using proprietary software. This was widely used from the first systems in the 1980s and means that some CP engineers have very old obsolete computers running obsolete, unsupported software in order to communicate with their older, still functioning CP systems. A detailed description of T/R power supply units and monitoring systems is given in Chess and Broomfield, 2014.

The number of systems an organization has in operation is one factor in choosing remote control. It becomes more cost effective to collect remotely as the number of structures increases. The sophistication of the client and his consultant or building/structure manager is another factor.

The reliability of the microprocessor system has not been reported in the technical press, although the author's first few remote control systems installed in 1986/1987 worked until the structures were demolished, most of those are installed inside buildings, in very benign environments. Some systems are comparatively simple and will only monitor on and off potentials, current and voltage. It is not possible to change the current or voltage settings on some of these systems. With modern microprocessors and the internet it is possible to store data on line, and send alarms by email or text message if the system malfunctions or exceeds defined limit values of current, voltage or reference electrode potential. Systems have been developed that will commission and operate a system according to BS EN ISO 12696. In their review of over 100 systems Polder *et al.* (2014) found that manual systems were more

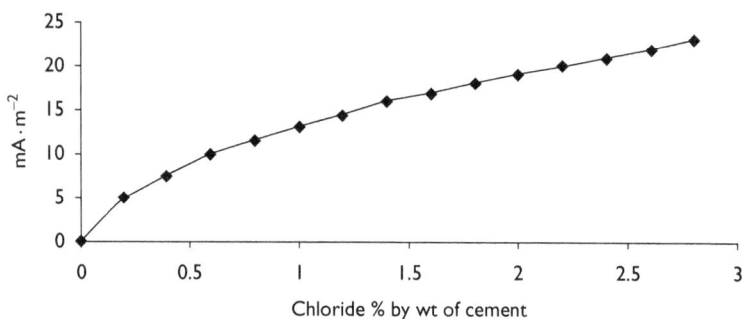

Figure 7.17 Cathodic protection current demand vs. chloride content at rebar depth (based on Bennett and Turk, 1994).

cost effective up until about 15 years life, at which point remote monitoring became more cost effective as the lower initial cost manual system becomes more expensive due to more frequent visits being required.

7.5.3 Monitoring probes

In order to measure the effect of the cathodic protection current on the reinforcement, probes are embedded in the concrete. The most common probe is the embeddable reference electrode or half cell. A number of formulations have been used but the most popular is the silver/silver chloride/potassium chloride (Ag/AgCl/KCl) reference electrode (Figure 7.16). A proprietary manganese/manganese dioxide electrode is also used extensively in Europe. Carbon, coated titanium and lead have been used as 'relative' or 'pseudo' references rather than 'absolute' references. They can be used for measuring potential shifts as long as they are stable over the period of measurement. A review of reference electrodes for concrete, their installation and relative potentials vs. the ASTM C876 criteria is given in NACE 11100 (2018). Conversion tables for the relative potentials of different reference electrodes are given in Annexe D of BS EN ISO 12696.

Due to the perceived low reliability of half cells in North America in the 1970s and 1980s alternative probes were developed. These include the current pick up probe. A section of steel is embedded in the concrete. This picks up a proportion of the current and is used to gauge the effectiveness of the system.

The steel probe is sometimes embedded in an excessively salty patch. With no current applied, a macrocell current flows from the probe to the reinforcement if they are connected with an ammeter between them. As CP current is applied, the current reduces and then reverses. This is called the macrocell or null probe approach and is used to show that a very anodic area has been made cathodic. It is therefore assumed that all of the rest of the steel is cathodic too. However it is dependent upon the amount of salt

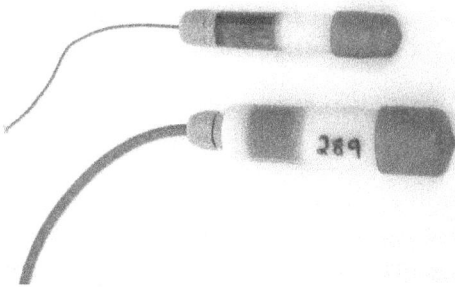

Figure 7.18 Two embeddable Ag/AgCl/KCl reference electrodes 15 mm and 25 mm diameter.

added to the patch, and if the salt diffuses away then it may no longer be the most anodic area after a few years.

An alternative is to identify the most anodic area of the zone and to isolate a short section of steel to form a macrocell probe without disturbing the concrete around the probe as discussed in SHRP-S-347 Bennett and Schue (1993) and Bartholomew *et al.* (1993). This is more realistic than embedding a probe in salty concrete but suffers the same problem with chloride movement with time.

7.5.4 Zone design

One frequently asked question concerning cathodic protection systems is 'what happens at the anode edge? Is there a risk of accelerated corrosion?' This is a valid question and the risk is supported by the Pourbaix diagram which shows areas of imperfect passivity, pitting and corrosion around the immune and passive regions (Figure 7.2). However, the author knows of no atmospherically exposed reinforced concrete structure that is totally protected by cathodic protection. Most have anode zones that end before the reinforced concrete does. No cases of accelerated corrosion have been reported between zones or at the end of zones.

It is rare for a cathodic protection system to consist of one continuous anode passing current from a single power supply. It is normal to divide the structure or elements to be protected into zones that are powered and controlled separately, and electrically separated by a gap of typically 25 mm.

Zone design can be dictated by requirements for different anode types (coatings on walls; mesh on floors probes in mortar joints around steel beams, etc.) They can be based on different current requirements, with high chloride areas divided into smaller zones to give each zone a fairly uniform current demand. They can be designed around different steel densities in different parts of beams or columns. In the United States, cathodic protection designs on decks have had zones in the 500–1,000 m^2 range, and bridge substructures 100–500 m^2 (AASHTO, 1993). This probably has more to do with geometry of deck slabs and substructures than with variations in local current demand. Steel framed masonry and brick clad structures are often designed with very small zones or sub zones.

7.6 Control criteria

At the beginning of Section 7.4 we saw that the T/R is typically designed to deliver 10 to 20 mA·m^{-2} of steel surface area. We have also seen that we must minimize the anodic reaction to reduce acid attack at the anode/concrete interface. This can reduce bond and thus increase the resistance of the circuit. It is therefore essential that we apply enough current to stop corrosion but only just enough. The control criteria for cathodic protection of atmospheri-cally exposed concrete has been the subject of considerable debate.

The appendix of BS EN 12696 states that the current demand may range from 2 to 20 mA·m^{-2}. This is a function of the chloride content at the steel and the amount of new concrete placed around the steel. In an investigation into these issues Broomfield (2004) found that the actual current density on a bridge beam was more accurately predicted by determining the chloride content at the reinforcement and using Figure 7.17.

The problem for the cathodic protection engineer is that the system must be designed and operated to achieve criteria in the national and international standards that demonstrate that corrosion has been suppressed. These cri-teria are not based on the current density on the steel but on the potential of the steel or potential shifts measured by embedded reference electrodes.

At the beginning of the section on cathodic protection it was explained that cathodic protection works by injecting so many electrons into the rein-forcing network that the anodic reaction cannot occur so corrosion stops.

The question is, how can we show that this has happened? In cathodic protection of steel in soil or water it is usual to do this by achieving a poten-tial of −770 mV or −850 mV against a copper/copper sulphate half cell on the surface as the system is switched off (the instant off potential). However, these criteria are not appropriate for steel in atmospherically exposed concrete for a number of theoretical and practical reasons. Two of the practical reasons are the difficulty in accurately measuring an absolute potential over a number of years when reference electrodes calibration may drift, and the fact that if an absolute minimum (or maximum negative) potential is achieved then some parts of the structure will be overprotected as the corrosion environment varies so rapidly and severely across a high resistance electrolyte like concrete.

In practice, the best control criterion is based on a potential shift. Theory and experiment tell us that a shift in potential of 100–150 mV will reduce the corrosion rate by at least an order of magnitude. Field evaluations have shown that this stops all further signs of corrosion damage in cathodically protected structures.

This has led to a number of criteria because of the practicalities of measuring potential shifts. The potential shift can be measured when switching on the system and switching it off. It must be measured as an 'instant off' potential when the system is on so that the CP current does not interfere.

Also the system fully polarizes and fully depolarizes very slowly. However, the potential is also susceptible to changes in the environment (temperature, humidity), as well as the applied current, so reading must be taken within a reasonable time period while the environment around the structure is reasonably stable. This is particularly important for tidal zones of structures in marine exposure conditions where the polarization/depolarization period should either be through a full tidal cycle or while the steel adjacent to the reference electrode is fully exposed. The full tidal cycle (about 12 h 20 min) is preferred as the steel should be in the same condition at the beginning and the end of the period while its condition will be changing (drying out) during an exposure period

Therefore the 100–150 mV shift in potential is often measured as a 100 mV shift from instant off to a period typically four hours later. This is measured at an actively corroding (anodic) location. Measurements are made throughout the four hour period so that the depolarization curve can be plotted. If depolarization continues at four hours then it is reasonable to expect that the 100–150 mV criterion has been achieved.

A number of things happen when cathodic protection is applied to steel in concrete. The negatively charged chloride ions move away from the negatively charged reinforcing steel and may be discharged at the anode and hydroxide ions are generated. Figure 7.19 shows how the chloride profile changes from an area subject to cathodic protection with the chlorides severely reduced at the cathode and also at the anode where they can be discharged as chlorine gas or may be washed away by surface water. The movement of chloride and the generation of hydroxyl ions suppress corrosion and rebuild the passive layer. This means that although the 100 mV shift is lower than the optimum 150 mV, the other factors are also mitigating in our favour (see Glass *et al.*, 2000 and references therein). During the initial few months of energizing and commissioning a cathodic protection system it is not necessary to instantly provide total protection as the ionic movements will build up with time. This is discussed further in the sections on energizing and commissioning CP systems.

The 100–150 mV criterion is straightforward to apply and is the most universally agreed criterion. Other control criteria such as the plotting of the applied current against the log of the potential (ElogI), absolute

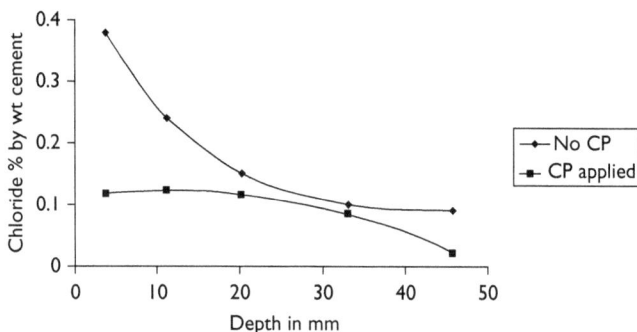

Figure 7.19 Comparison of chloride profiles in CP and non CP areas Yaqina Bay Bridge soffit based on Broomfield and Tinnea, 1992.

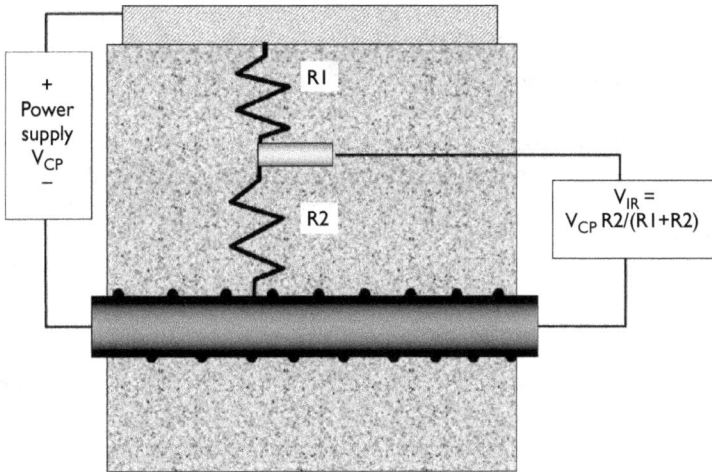

Figure 7.20 Two-dimensional representation of the iR drop through concrete. While the CP current is flowing, the voltmeter between the reference electrode and the steel measures the steel/concrete potential plus the iR drop V_{IR} which is a function of the CP voltage V_{CP} and the resistances between anode, reference electrode and steel.

potentials, macrocell or null probe current reversals and other potential shifts have been used and are used by some cathodic protection specialists but there is some controversy about their theory and practice.

7.6.1 The iR drop and instant off measurement

The iR drop is the additional voltage seen between the reference electrode and the steel because a current is passing through concrete with a significant resistance. Figure 7.20 shows a two-dimensional illustration of how this voltage occurs when the resistance from the anode to the reference electrode is R1 ohms and the resistance from the reference electrode to the steel is R2 ohms. If the applied cathodic protection voltage is V_{CP} then the potential measurement between reference electrode and steel has an additional iR component.

7.7 Standards and guidance documents for cathodic protection of steel in concrete

The European Committee for Standards (CEN) has a single standard for steel in concrete. BS EN ISO 12696. This has been adopted by the International Standards Organisation. This covers impressed current and galvanic cathodic protection for atmospherically exposed, buried and sub-merged conditions. NACE/AMPP has a series of standard practice docu-ments, test methods and technical reports covering these areas separately as listed in Table 7.2.

A review of the performance and acceptance criteria for cathodic protec-tion of steel in concrete has been published by Broomfield (2020) and a historical review of ICCP of steel in concrete has also been published by Broomfield (2021).

7.7.1 AMPP NACE standards

The AMPP NACE standard practice SP0290 was the first to be produced (1990) and has been revised since the original. Its scope states that it excludes prestressed concrete. There is a separate NACE report on cathodic protection of prestress NACE 01102. This gives valuable information on applying CP to prestressing but there has been no pressure to convert this to a standard.

NACE SP290 It is a shorter document than BS EN ISO 12696 and is specifically concerned with control criteria which are at the front of the document in Section 2. The first criterion mentioned is 100 mV polarization development or decay to be measured between the 'rest', 'equilibrium', or 'natural corrosion' potential and the instant off potential measured 0.1–1.0 seconds after switching off the system. The instant off potential is required to remove the 'iR drop'

Measurements are to be taken and the 100 mV criterion achieved at the most anodic location in each 50 m^2 area or zone or at artificially constructed anodic sites . . . provided its decayed potential or decayed off potential is more negative than −200 mV vs. copper/copper sulfate reference electrode CSE.

This perhaps shows the origins of this standard which was written for US and Canadian highway bridge decks which is where the 50 m^2 zone and −200 mV (ASTM C876) figures come from. A degree of interpretation would be required for a building with many small zones of 5–10 m^2 and with carbonated concrete where corrosion might occur at potentials more positive than −200 mV CSE in disagreement with ASTM C876.

There is also discussion of the E LogI criterion where a graph of potential versus the logarithm of the applied current is plotted. The transition of the steel from anodic to cathodic condition is indicated by a break in the curve. However, the existence of a break and its location can be subjective and the shape of the curve changes with the scan rate. The RP suggests that the method is used to determine the initial cathodic protection current requirement.

The standard states that the maximum sustained current density between the anode and the concrete should not exceed 108 mA/m^2 which is correct for the anode types described but is exceeded in the manufacturers' data sheets for many probe anode designs (see Section 7.3.6) which allow far higher current densities.

As stated in the anode descriptions earlier, there are also two NACE test methods for cathodic protection anodes. These are TM 0294 on embeddable anodes (mixed metal oxide coated titanium, mesh, ribbon, tubes, rods and conductive ceramic tubes) adopted in an edited form by ISO as ISO 19097-1 and TM01105-2005 on carbon pigmented conductive coating anodes. In addition, there is a specification for applying thermal sprayed zinc anodes to concrete American Welding Society (2002).

At the time of writing, AMPP is developing a new standard exclusively on criteria to be applied to all the AMPP standards that concern CP of steel in concrete: AMPP SP21520 Acceptance Criteria for Cathodic Protection of Steel in Concrete Structures. This means that criteria will be removed from other standard practice documents to avoid variations between standards as they are revised at different times.

7.7.2 CEN Standard BS EN 12696

The CEN standard states in its scope that it refers to atmospherically exposed normal reinforced and prestressed concrete as uncoated and organic coated reinforcement. It is more detailed than the AMPP/NACE standard (65 pages including annexes vs. 15) with far more on the assessment and repair of the structure and installation procedures. The criteria are buried in Section 8 of the ISO standard.

There are very useful informative annexes on the principles of cathodic protection in concrete, the design process, with recommendations for design current densities discussed earlier in this section, and a brief description of anode materials and systems.

The criteria (Section 8.6 of the standard) start with a requirement that no (instant off) potential should exceed a limit of $-1,100$ mV with respect to Ag/AgCl/0.5 M KCl for reinforced concrete and -900 mV for prestressed concrete. This is aimed at minimizing hydrogen evolution with a wider margin for prestressed steel where the consequences could be more extreme as discussed in Section 7.2.1.

Having defined an absolute limit on potential, the engineer is then offered a choice of any one of three criteria to choose from to be met at 'any representative point':

- An instant off potential more negative than -720 mV vs. Ag/AgCl/0.5 M KCl – this criterion is the conventional one for buried and submerged structures.
- A potential decay over a maximum of 24 h of at least 100 mV from instant off – this is very similar to the NACE depolarization criterion but has a time period specified.

- A potential decay over a period longer than 24 h of 150 mV – this was discussed at the beginning of Section 7.4 and is based on the work of Bennett and Broomfield (1997).

7.8 System installation

The installation of the different components of the cathodic protection system is described later. The order given is logical but not necessarily the one used on site where work may proceed in parallel or a different sequence to fit in with other requirements of site work.

7.8.1 Patching for cathodic protection

Patching for cathodic protection is merely to ensure that there is ionic continuity between the steel and the anode (Figure 7.1). This means that the patch repair material must have the following properties:

1 It must conduct ionically not electronically. There must be no conductive filler such as carbon or a metal. Zinc is used in some patch materials to minimize the incipient anode effect, they must not be used for any electrochemical process repairs. Metal or carbon fibres must not be used either.
2 The resistivity of the patch material must be low enough to allow current to pass into the steel. A resistivity of 15 kΩ·cm at 28 days under saturated conditions has been specified by the United Kingdom DoT based on successful use of materials of this resistivity on the Midland Links. There is some disagreement about this specification but it is the only quantitative information available. In the United States some applicators use salt in the concrete to 'even out' the resistivity with the parent concrete. This practice is rarely accepted elsewhere. BS EN ISO 12696 advises that the concrete repair material should be within 50–200% of the nominal parent concrete electrical resistivity. However, there are caveats regarding measurement of resistivity and changes with time.

Some practitioners strongly feel that the quality of the patch repair is far more important than the relative resistivities of the parent concrete and patch repair material and that as long as metallic conductors (including carbon) and epoxy concretes are excluded repairs will do their job while cathodic protection does its job. A discussion of the issues relating to the resistivity of patch repair materials is given in Broomfield (2018a).

Patching for CP only requires the removal of the corrosion damaged cover, the cleaning of the rebar and its reinstatement. The lack of structural implications and the comparative cheapness of the concrete patching requirements are two important factors in choosing CP for repair. It must be noted that good quality, low shrinkage materials must be used and well applied. CP will not stop a badly applied patch from falling off or developing cracks if badly cured. The author is aware of plastic shrinkage problems in repairs for cathodic protection projects but not of corrosion problems.

Excavated areas may be filled with the same material used to apply the anode, for example, sprayed concrete on substructures or the concrete overlay on decks. In the case of the titanium mesh special fixings will hold the anode in position as the concrete fills around it.

7.8.2 Rebar connections, electrical continuity of reinforcement and stray currents

Electrical connections may be made by self tapping screws, welding or brazing or other techniques that provide a durable low resistance metal to metal bond. Extra bars may need to be welded in if continuity between bars is inadequate. Connections are usually protected with epoxy glue or putty applied over them. Cabling may be run through the concrete overlay, over the anode or from the back of the protected surface. Cables are run into junction boxes for splicing and connections. There must be at least two connections per zone for redundancy. There may be separate connections for the reference electrodes.

Rebar continuity is essential to avoid stray currents that can accelerate corrosion. Figure 7.21 shows how an isolated rebar between the anode and the cathode will be cathodic where the current enters the steel and anodic where it exits. This will accelerate corrosion at the anodic site. Although there are few serious cases identified in cathodic protection systems, this is a greater concern for realkalization and desalination systems where the charge density is higher.

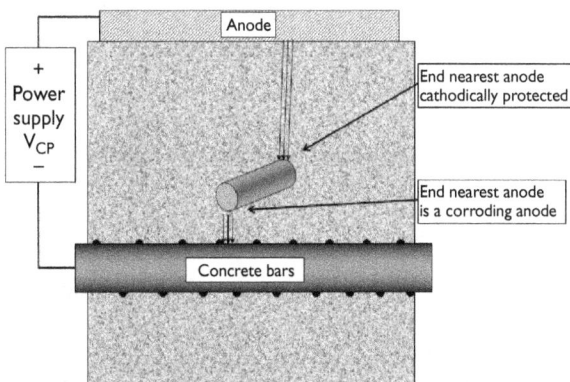

Figure 7.21 Illustration of how an isolated piece of steel can be subjected to stray-current-induced corrosion, as the current enters at the end nearest the anode cathodically protecting it, and leaves at the end nearest the cathodically protected steel, causing corrosion.

There is a CPA Technical paper on stray currents (CPA Technical Note 10) as well as a NACE/AMPP standard practice SP 21427 and a technical report TR01110. Stray currents can arise from the effect of a cathodic protection system on isolated steel in the protected structure. This can be trivial such as nails or tie wire left in the concrete during casting. It can also be more significant lengths of rebar in lightly reinforced structures where corrosion of tie wire is significant. The source of stray current can be external such as from DC traction systems (trams, street cars, light rail). There is evidence from cathodic protection of pipelines that alternating current can affect cathodic protection systems by moving the potential to less negative values and therefore not providing adequate protection to the steel.

Generally there is far less evidence of stray-current-induced corrosion from external sources into atmospherically exposed reinforced concrete structures compared to buried structures generally and far less evidence of stray-current-induced corrosion in buried or submerged concrete compared to buried or submerged steel, as concrete has much higher electrical resistance. However, it can occur and can be tested for in the same manner as for other metalwork. There is discussion of interference testing and limits for stray currents in CPA Technical Note 10 (Atkins 2018) and in BS EN 50162:2004.

7.8.3 Monitoring probe installation

Reference half-cell electrodes should be installed in anodic sites without disturbing the concrete around the steel to be measured. A method of installation is shown in NACE 11100 (2018). There should be at least one electrode per zone and usually more. Other probes such as macrocell probes, null probes or electrical resistance probes (Section 5.2.4) may also be installed but usually in addition to half cells. For systems designed to minimize future access then pseudo reference electrodes are frequently installed along side true reference electrodes. As true references have a finite life (typically 20 years or so), the pseudo references, also known as decay probes, can be used without having to replace true references when they reach the end of life.

7.8.4 Anode installation

Anode installation should be done according to the manufacturer's instructions. The anode must not come in direct electrical contact with the reinforcing steel. There must be no metallic fixings into the anode.

The anode system may be a single component, such as flame sprayed zinc, or multiple component such as a titanium mesh with a cementitious overlay. It may consist of a single continuous anode such as mesh or coatings, or multiple anodes connected together such as ribbon anodes or the discrete rod anodes. All anodes require electrical connections to the power supply. As for the rebar connections these should be duplicated for redundancy.

The size and shape of the anode zone will be determined by a number of factors including:

- Steel surface area density and layout (beam soffits have higher steel density than sides so may be separate zones).
- Exposure conditions (tidal, splash and above splash zones may require separate zones).
- Different elements (beams versus decks even when beams are integral).
- Uniform current distribution (minimizing the current range from the largest to the smallest zone).

In some cases sub zones with adjustable resistors are used to subdivide a zone without multiplying up the total number of current outputs. The number of zones can range from a single zone on a small structure to over 100 zones on large complex structures.

7.8.5 Transformer rectifier and control system installation

The process of running mains power, installing the T/R and control system and wiring up can be done as a single operation or in stages. All probes and leads should be tested before being connected to junction boxes and again before connection to the T/R and control system and then tested once the system is installed. T/Rs are factory tested and must comply with national electrical regulations and codes for safe and efficient operation.

Lightning arrestors may be required on some atmospherically exposed structures. This is a complex subject and some owners seem to expect them to prevent any damage to the CP system. In fact the lightning arrestor circuit will frequently be destroyed but the rest of the electronics survives which means the circuit has done its job. The existence of a large sheet of conducting material (the anode) will make the cathodic protection system vulnerable to lightning strikes and the performance of building lightning conductor/ arrestor systems should be checked on a structure known to be vulnerable to lightning strikes.

7.8.6 Initial energizing

Anode durability can be strongly affected by the initial energizing and commissioning process. This is particularly true of conductive coating systems where the anode/concrete bond is crucial to the long-term durability of the system. Passing too much current too quickly through the anode can generate large amounts of acid at the anode as the chloride level is highest there if chlorides are being transported into the concrete from the outside. Acids are formed by reaction of the chlorides to form chlorine gas, hydrochloric and hypochlorous acid. This will attack the concrete paste, consume the anode and reduce the system lifetime, particularly for coating anodes. The problem is less important for mesh and ribbon type anodes. It is therefore recommended to energize the system as gently as possible consistent with demonstrating that the system is fully effective and provides protection. Another advantage of slow energization is that the lowest current level consistent with adequate protection can be found while turning the current up too high initially can make it very difficult to find the optimum current.

A series of checks should be carried out to ensure that all anodes and probes are correctly wired. Energizing should be at very low current levels initially and should be carried out after the anode system is fully cured and ready according to the manufacturer's instructions.

7.8.7 Commissioning

A series of commissioning tests should be agreed at the design stage. These may be taken from the AMPP/NACE SP 0290 or BS EN ISO 12696 standards. However, the issue of anode durability is still important. A slow increase in current to achieve the minimum required control criteria is recommended. As chlorides move away from the rebar under the influence of the electric field, it is better to underprotect in the first months rather than over protect. As long as the system is seen to be capable of delivering adequate current and all test and monitoring areas are polarizing in a sensible manner, with all anodes passing current fairly uniformly, the system should be accepted as functioning correctly.

The AMPP NACE SP0290 100 mV polarization criterion is a very useful start up criterion to avoid initial overprotection.

7.8.8 Operation and maintenance

It is essential that impressed current cathodic protection systems are properly maintained. This usually means that a budget system of training for personnel and maintenance schedule is developed by the client, engineer, corrosion specialist and contractor at the initial design and specification stage.

The client may undertake maintenance himself or have an agent do it for him. Even with remote control and monitoring a visual check on the system should be undertaken every month or so and an annual inspection conducted leading to repair and maintenance if required.

7.9 Cathodic protection of prestressed concrete

Extreme caution must be used when applying cathodic protection to prestressing steel or to elements that include prestressing steel. This is for two reasons:

1   Because of the risk that if the potential exceeds the hydrogen evolution potential hydrogen embrittlement could occur with potentially catastrophic failure of the steel.
2   Because if the prestressing steel has corrosion pits in it, those pits are potential stress concentrators and there is a risk of either under protection in the bottom of the pit which could lead to continued corrosion and then to failure of the stressed strands, or of overprotection leading to hydrogen embrittlement.

These issues are extremely well reviewed in NACE report 01102. The report includes 'qualification criteria' for applying cathodic protection to prestressing which depends upon susceptibility to hydrogen embrittlement, checks on the degree of section loss and pitting. It is made clear in the report that if cathodic protection is to be applied then the preference is to use a galvanic system as the risk of exceeding the hydrogen evolution potential is negligible. If impressed current cathodic protection is to be used then the condition of the steel and the control criteria are of paramount importance.

The CEN standard BS EN 12696 gives criteria for controlling cathodic protection systems applied to prestressing steel by keeping the potential more positive than -900 mV with respect to Ag/AgCl/0.5M KCl. NACE RP0290 states in its scope that it does not apply to prestressed concrete structures. NACE RP 0100–2000 gives a limit of -1000 mV with respect to CSE (approximately -900 mV with respect to Ag/AgCl/0.5M KCl).

In Italy CP has been applied to new bridges to protect anchorages, reinforcing steel and exposed tendons in prestressed, post-tensioned bridges (Baldo *et al.*, 1991). The important factor here is that cathodic protection is applied from construction and therefore to uncorroded steel of low risk of hydrogen embrittlement. The Italian approach is conservative as very modest currents and potentials are needed to protect new structures. However, applying CP to a new, undamaged structure is thought to admit poor confidence in the initial design by many engineers unless the structure is exposed to a very severe environment such as salt or brine containers. The Italian choice was based on a lowest maintenance option as closing lanes on elevated highways would require complex expensive traffic control, high risk to site operatives and heavy traffic congestion.

In the United States trials of impressed current cathodic protection have been conducted on pre-tensioned structures (Bennett and Schue, 1998).

Many pre-tensioned bridge piles have been cathodically protected in Florida using galvanic anodes. This is unlikely to cause overprotection and hydrogen evolution.

It has already been noted that a United Kingdom post-tensioned bridge collapsed in Wales a few years ago. The bridge had no waterproofing on the deck, the ducts were poorly grouted and salt got in between segments directly onto the steel. The UKDoT has undertaken an extensive survey to determine how many bridges are in a comparable condition in the UK. As CP cannot protect the cables inside ducts on post-tensioned structures, it has limited applicability for protecting such structures. The definition of a susceptible steel that needs to be considered at risk has received much attention in recent years. BS EN ISO 12696 states that high yield steels >550 N/mm^2 should be investigated.

Thus it can be seen that each prestressed structure must be evaluated to determine the feasibility of cathodic protection current reaching the steel that needs protection, whether galvanic cathodic protection can be applied and if not using qualification criteria such as those in NACE 01102 (2002) to determine if the structure is suitable for impressed current cathodic protection.

7.10   Cathodic protection of epoxy coated reinforcing steel

Because of the problems of corrosion of fusion bonded epoxy coated reinforcing steel (FBECR) in Florida (Clear *et al.*, 1995; Manning, 1995), the Florida DoT and other US researchers have investigated methods of cathodically protecting FBECR. The main problems are establishing the continuity of the steel and the risk of pitting and undercutting corrosion behind the epoxy coating.

A survey of the FBECR structures in Florida that have suffered from corrosion has shown that many have a very high level of continuity. On a survey of 10 bridges, there was an average of 27% readings showing electrical continuity (Sagüés, 1994). This may be due to the use of uncoated tie wire in early construction work cutting through the coatings and the effect of squeezing bars together at ties.

In order to install cathodic protection, continuity can be established by welding in extra rebars. However, at Florida DOT one approach has been to expose bars in damaged areas, grit blast them clean and apply arc sprayed zinc directly onto the steel and then across the steel surface. This provides galvanizing directly on the steel and SACP to the steel embedded in the concrete. Multiple continuity connections are established by the sprayed zinc.

The problem of pitting and under coating corrosion is more difficult and is well known in pipeline corrosion where fusion bonded epoxy coatings are frequently applied to the outside of pipelines and then cathodic protection applied to protect the pinholes that inevitably occur. However, these systems are applied from new so there is no corrosion established. The FBECR structures are already corroding when CP is applied. It is therefore possible for corrosion to be established under the coating where the cathodic protection current cannot reach. The small driving voltage of a galvanic anode system means even less protection or penetration of current than for an impressed current system. In practice the risks of lack of continuity and of under-film corrosion must be accepted. The galvanic sprayed systems installed on prestressed concrete piles with FBECR have lasted 10–12 years with no signs of further corrosion.

7.11   Cathodic protection of structures with ASR

Alkali-silica reactivity (ASR) is a condition where certain silicateous aggregates are susceptible to the alkalinity in the concrete and react to form silica gel. Silica gel is used as a desiccant because it absorbs moisture. As it absorbs moisture it swells. ASR is characterized by 'map cracking', and weeping of the gel as a white efflorescence from the cracks (BRE Digest 330, 1999). One way of controlling ASR at the mix design stage is to limit the maximum alkali content of the mix. This can be done by controlling the chemistry of the cement powder or by blending the cement. In these cases marginally susceptible aggregates can be used.

It has been shown earlier that cathodic protection creates hydroxyl ions and will also attract positive ions such as sodium and potassium to the steel. This will increase the alkalinity around the reinforcing bar. In principle this could cause ASR or accelerate ASR in susceptible mixes. This has been demonstrated in the laboratory. However, there are no recorded cases of ASR being caused or accelerated by CP in field structures. In a review of field structures by SHRP, a structure with ASR showed no acceleration in the ASR in areas where CP was applied. The issue is also discussed in Mietz (1998). In the standard on cathodic protection BS EN ISO 12696 Appendix A.6 states that 'cathodic protection applied in accordance with this document has been demonstrated to have no influence on alkali silica reaction/alkali aggregate reaction (ASR/AAR).'

7.12   Electrochemical chloride extraction

We have already determined that the chloride ion is a catalyst to corrosion (Section 3.2.3). As it is negatively charged we can use the electrochemical process to repel the chloride ion from the steel surface and move it towards an external anode. This process, called electrochemical chloride extraction (ECE), desalination or chloride removal, uses a temporary anode and a higher electrical power density than CP, but is otherwise similar (Figure 7.1).

Preparation in terms of concrete repair, power supplies etc are similar to those for impressed current cathodic protection except that the power supply is temporary and may be from a temporary source such as a generator. The output is higher, up to 40 V and 2 A·m². Standards and guidance documents for ECE and realkalization are discussed in Section 7.13.

7.12.1 Anode types

The most popular anode is the same coated titanium mesh used for impressed current cathodic protection. Instead of embedding it permanently in a cementitious overlay a temporary anode system is used. A proprietary system developed in Norway consists of shredded paper and water sprayed onto the surface to form a wet 'papier mâché'. The mesh anode is then fixed to the surface on wooden batons and a final layer of papier mâché applied. The system is kept wet for the operational period. Figure 7.22 shows an early installation underway using a mild steel mesh anode. This is rarely used now for ECE.

Figure 7.22 Electrochemical chloride extraction being applied to the Burlington Skyway Ontario, Canada 1989, using cellulose fibre over a mild steel mesh. Courtesy De Nora Group.

An alternative to the cellulose fibre system is a wetting system of water baths or ponds which have been applied on bridge decks in the United States. An electrolyte solution circulating system on vertical surfaces using 'blankets' has been used in the SHRP trials on bridges in Florida, Ohio and New York and Ontario (Figure 7.23). A system of surface mounted tanks or 'cassettes' has been developed in the United Kingdom as shown in Figure 7.24. This can contain mixed metal oxide coated titanium mesh or mesh ribbon anode.

7.12.2 Electrolytes

The electrolyte is the liquid in the cassette, cellulose fibre, blankets, ponds or other containment system that transport the current from the anode to the steel. Potable water is a suitable electrolyte but is liable to acidify and promote the evolution of chlorine gas. Regular replacement will be required to maintain the pH at a suitable level. Concrete suffers serious acid attack below pH 5 so the pH should not be allowed to drop below pH 6. Alkaline electrolytes are generally beneficial to the performance of the system, to minimize the risk of etching of the concrete surface and to minimize chlorine gas evolution on inert anodes. The following alkaline solutions are suitable.

- Saturated calcium hydroxide with an excess of solid material to maintain the saturation.
- Lithium borate is suitable where there is concern about alkali-silica reaction in the concrete.
- Plain water is preferred by some designers to maximize chloride transport.

Figure 7.23 Geotextile blanket anode system being installed on a bridge column in the SHRP ECE trials (Bennett *et al.*, 1993a, Acknowledgement Transportation Research Board).

Figure 7.24 Cassette type anode attached to a bridge abutment. Courtesy Makers Ltd.

A proprietary system, extensively used in Germany, uses an ion exchange system to regenerate the electrolyte (Schneck, 2006; Schneck, 2016).

7.12.3   Operating conditions

The very first trials of this system were based on a rapid treatment period of 12–24 hours. Trials were done in Ohio, United States, and lab tests in Kansas Department of Transportation in 1976 (Morrison *et al.*, 1976, Slater *et al.*, 1976). However, a more fundamental study carried out by the Strategic Highway Research Program (SHRP) in 1987–1992 (Bennett *et al.*, 1993a,b) showed dangers in applying more than 4 $A·m^{-2}$ of steel or concrete surface area. Current treatment times are now measured in weeks rather than days.

Ontario Ministry of Transportation carried out a trial of the Norwegian system on Burlington Skyway in 1989. The voltage was kept at about 40 V. The total charge passed was 610 $A·h·m^{-2}$ of concrete surface area over 55 days giving an average current density of 0.462 $A·m^{-2}$. This structure had a low steel concentration so removal was patchy, very high over the rebars but lower between the bars. They found 78–87% of the chloride removed directly above the rebars and 42–77% of the chloride removed between the rebars.

A trial was carried out in the United Kingdom on a section of crosshead taken from the corroding substructure of a bridge. The section was removed to a contractor's depot and the sprayed cellulose fibre system applied. The system ran for 92 days passing a total of about 19,565 A·h charge through approximately 11 m² of steel surface. This gave a charge density of 1,704 A·h·m^{-2}, an average current density of 0.77 A·m^{-2} and a power density of 25 W·m^{-2}.

This work was followed up by laboratory tests on concrete to rebar pull out strength and is discussed later.

A typical system will run at no more than 4 A·m^{-2} of steel surface area for a period of 2–8 weeks.

The following information was supplied by a practitioner with over 20 years of applying ECE (Schneck, 2021).

- ECE can be considered to be a safe and reliable non-destructive rehabilitation method; the only (but catastrophic) threat is connecting the wrong polarity between the anode and reinforcement. Even in difficult circumstances – high chloride ingress and low concrete cover – with repeated treatment cycles corrosion activity can be eliminated durably. However, there can be economic limitations that make more than one or two repetition stages impractical.
- The process of ECE is self-regulating. At a constant voltage that should be about 40 V for establishing a strong electrical field within safety limits, the local electrolyte and polarization resistances influence the flowing current. At low resistances very high currents can be measured initially (up to 20 A/m² reinforcement); they drop usually within one week if no short circuit is the cause. It is not advised to run ECE in a constant current mode and to avoid high currents at all – this limits the electrical field and the desalination effect, and although chloride migration and impressed current correlate, they do not depend on each other.
- The water in the concrete pores plays an important role for the ECE. Much water is moved into the concrete by the electrical field from the wet anode array of ECE (the water content on the concrete cover zone can be raised from 3% to more than 6% within a few weeks), and the capillary movement of the evaporating water can 'lift' chloride from behind the outer rebar layer into the cover zone during a break between treatment stages.
- The effect of corrosion protection is achieved from both chloride removal and concrete alkalisation. At a good charge input (>400 to 1,000 Ah/m²), some residual chloride – even up to 1% in rebar vicinity – will not block a solid re-passivation of the reinforcement. Repeated corrosion surveys on concrete parts showed even 10 years after an ECE

treatment potential being increasingly positive – a long time after any surplus water has been evaporated.

- Enhanced corrosion measurement methods (AC impedance, galvanostatic pulse, linear and Tafel polarization) can give instant and plausible insight in the ECE rehabilitation progress, even at heavily disturbed potentials. It can be seen how resistances go up considerably and corrosion currents drop, and as long as the resistances stay low, ECE should be continued. Also the current feed is a useful indicator of treatment progress – as long as the currents are higher than 1 A/m², it's no time yet for termination in most cases. If the currents drop below 0.5 A/m² from much higher values before, the end of treatment is likely.

7.12.4   End point determination

There are now two major standard documents and a guidance document on ECE. These define suitable end points for treatments. BS EN 14038-2 (2021) includes a UK annexe which supplements the main text. It also refers to a testing method to determine the operating parameters, particularly time of treatment in AMPP/NACE Technical Report TR01101(2018). That report was the basis for the NACE/AMPP standard practice document SP0107 (2021).

End point determination can be by several means:

i   Point of diminishing returns – resistance goes up, amount of chloride removed goes down, when the current is small and the amount of chloride being removed has diminished, switch off. Switching off for about a week will bring the system resistance down. How much more chloride is removed by allowing 'rest' periods is a function of a number of parameters, which are not well understood.

ii   Direct measurement – take samples from the concrete and when an agreed level is reached, stop. This may be a chloride level (typically less than 0.4% chloride by mass of cement within 25 mm of the reinforcement); or a chloride/hydroxyl ratio of less than 0.6 (see Section 3.2.3). This assumes that good sampling is possible and that samples are representative. A method of determining the chloride/hydroxyl ratio of concrete pore water is given in Cáseres *et al.* (2006).

iii   Indirect measurement – sample the anode system or electrolyte. When chloride level is either at a plateau or an agreed level, stop. This assumes good sampling.

iv   Experience of charge density needed – measure charge passed (amp hours per square metre) and when an agreed limit is reached, switch off. The present consensus is that the steel current density should be a minimum of 600 A·h·m^{-2} and should not exceed 1,500 A·h·m^{-2}

based on Bennett and Schue (1993) and many subsequent applications of ECE.

Sohanghpurwala (2003) re-evaluated the specimens and structures treated in the SHRP work (Bennett and Schue, 1993; Bennett et al., 1993a) and found the onset of corrosion after 10.25 years on specimens that had been treated at 600 A·h·m^{-2}.

In practice a combination of criteria are used. Experienced engineers and contractors will know the charge density needed and a trial may give a more definite value for a given structure (or element within the structure). Sampling directly and indirectly will show that the system is responding in the expected manner. The point of diminishing returns should be reached soon after the other thresholds.

It is impossible to remove all the chlorides from the concrete by electrical means. The area immediately around the rebar is left almost chloride free but further away there is less effect. This is particularly true behind the steel and for widely spaced bars. Chloride removal will deplete the amount of chloride immediately in contact with the steel and will replenish the passive layer. Somewhere between 50 and 90% of the chlorides are removed. Field data show that this is effective for at least 15 years but for how much longer is uncertain. The results now suggest a minimum of 15 years if ECE is well applied and subsequent ingress of chloride is controlled. The Burlington Skyway system was over 17 years old and showed no sign of corrosion re-initiating despite the fact that the chloride content between bars remains high (Pianca et al., 2003 and private correspondence 2006).

If large amounts of chloride have penetrated beyond the steel or were cast uniformly into the concrete, then chloride removal will mainly affect the chloride level in the cover concrete. However, treatment of structures with widely spaced bars has found more removal behind bars than where bars are closely spaced. The large reserves of chlorides in the bulk of the concrete may then diffuse back around the steel and the removal process may be very short lived. SHRP experiments on a marine substructure were discouraging (Bennett et al., 1993a). UK trials on marine piles demonstrated similar problems although a system on the deck soffit of the wharf was successful as it stayed out of the water (Armstrong and Grantham, 1996). The problem is leakage of current outside the anode system and leakage of chloride into it, leading to the electrolysis of sea water rather than chloride removal.

However, research shows that the most important aspect of the chloride removal process is the generation of hydroxyl ions, the rebuilding of the protective passive film, infilling of voids and the steel/concrete interface and the removal of the chlorides immediately around the rebar. SHRP research showed that even with a modest total charge passed and only 50–80% chloride removal, and with chloride levels still above the corrosion threshold, treatment will give a very low corrosion rate and very passive half-cell potentials which last more than five years without reactivation (Bennett et al., 1993c). If this is true then we should perhaps call the process 'electrochemical chloride mitigation' and avoid requirements to remove more than 90% of the chloride, as this may not be necessary. Glass et al. (2003) have shown how important these effects are.

7.12.5   Possible effects

Passing large amounts of electricity through concrete can have effects upon its chemistry and therefore its physical condition. Brown staining around the rebar has been observed on specimens when high currents (in excess of 4 $A\cdot m^{-2}$) are used. This is an effect on the concrete, not the steel. Current levels are therefore maintained at less than 2 $A\cdot m^{-2}$ (usually in the range 0.5–1 $A\cdot m^{-2}$).

There are two known side effects of ECE. The first is the acceleration of alkali-silica reactivity (ASR) and the other is reduction in bond at the steel concrete interface.

A third issue is hydrogen evolution. This is inevitable with ECE and so this process must not be used where current can flow to high strength steel under tension or prestressing steel.

7.12.6   Alkali-silica reactivity

Research at Aston University in England, and by Eltech Research in the United States under the SHRP program shows that ASR can be induced by the cathodic reactions ((7.2) and (7.3)) that generate excess alkali at the steel surface. This is exacerbated by the movement of alkali metal ions (Na^+ and K^+) to the steel surface under the influence of its negative charge. Some researchers in Japan have suggested that the pH can be so high that the silica gel dissolves, stopping the expansive process (Sergi and Page, 1992).

SHRP undertook a field trial in collaboration with Ontario to see if ASR can be controlled by the application of lithium ions in the electrolyte. These ions (Li^+) move towards the rebar under the influence of the ECE electric field. Lithium is known to reduce or stop ASR and has proved effective in lab tests. If a corroding structure is made with aggregates susceptible to ASR, a detailed investigation of its likely reaction to ECE will be required. There have been no published reports of its current condition.

It has been suggested that if a structure is suffering from ASR and corrosion then corrosion should be the first priority as corrosion will cause the most damage most rapidly. That decision can only be made on a case by case basis.

7.12.7   Bond strength

The effect of current on bond strength of steel in concrete has been a subject for discussion in the technical literature for many years. This is usually with

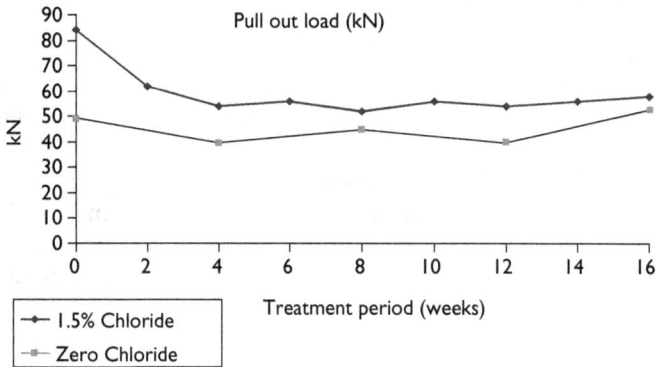

Figure 7.25 Maximum pull out load vs. time at 0.75A m^{-2} based on Buenfeld and Broomfield, 1994.

reference to cathodic protection but no effects have been observed in the field on the many hundreds of CP systems in service for up to 20 years now (Virmani and Jones, 1988). In most practical applications the major part of the bond is supplied by the ribbing on the bars so the details of the performance of the steel/concrete interface is irrelevant.

A laboratory study was undertaken to investigate the effect of pull out strength for the application of ECE to the Tees Viaduct, a long bridge structure in the North of England (Buenfeld and Broomfield, 1994). Smooth rebars were used in its construction due to steel shortages at the time of construction. Some earlier, unpublished tests had shown that ECE can reduce bond strength by as much as 50%. However, careful review of the laboratory data showed that a large amount of charge was passed to get this drop in bond: about five times as much as is used normally in an ECE treatment. A very high current density was also used. Figure 7.23 shows the effect of different levels of charge (current × time) on the pull out strength of smooth bars from cylinder specimens.

These experiments show that bond strength increases as the specimen corrodes (effectively prestressing the concrete by expansive oxide formation). The ECE current appears to eliminate this prestressing effect, although the pull out strength does not fall below the level of control (Figure 7.23).

If this is in fact the mechanism then the long-term effect must be considered. If corrosion has occurred over a period of 5–10 years, then the concrete may creep to accommodate some of the stresses generated by corrosion. If the ECE current then removes the stress it will not be possible for the concrete to revert elastically. Therefore there is a risk of a plane or cylinder of weakness developing around the bar, leading to a reduction of pull out strength.

7.12.8   Results after treatment: beneficial effects of passing currents through concrete

Comparisons of corrosion rates before and after treatment have consistently showed significant reductions. In the Tees Viaduct field measurements of corrosion rates were in the range $0.33–1.66$ $\mu A \cdot cm^{-2}$ with a mean of 0.377 $\mu A \cdot cm^{-2}$. On a block that was treated with ECE, a year later readings ranged from 0.0005 to 0.094 with a mean of 0.0028 $\mu A \cdot cm^{-2}$. Although no direct 'before and after' measurements were made at the same locations, this shows a two order of magnitude difference in corrosion rate between treated and untreated steel in concrete (Broomfield, 1995).

Resistivity measurement showed large increases after treatment, to over 200 kΩ cm in the field and 5 kΩ cm on untreated lab specimens to 30 kΩ cm after treatment on specimens vacuum saturated with distilled water. Petrographic microscopy showed the treatment to block pores, especially close to the steel reducing transport of water, oxygen and chloride ions. This is possibly due to redistribution of calcium hydroxide. There was also an improvement in freeze/thaw resistance (Buenfeld and Broomfield, 1994).

ECE can be considered a safe and reliable non-destructive rehabilitation treatment if carried out by a qualified and experienced practitioner. The most difficult conditions are presented when there is high chloride and low cover to the steel. This may require repeated treatments. Unfortunately, this may prove uneconomic. In most cases, once the current density drops below $0.5 A/m^2$ the treatment has been effective.

Experience shows that ECE is self-regulating. At a constant voltage of about $40V$ DC, the current density may start very high but then drop within a week. Successful treatment is usually achieved with a charge density of 400 to 1000 Ah/m^2, even if this leaves some chloride near the rebar. It has been found that this achieves very long term repassivation of the bars with positive reference electrode potentials 10 years after treatment (Ulrich, 2021). Electrochemical measurement techniques such as those described by Schneck (2019) can be useful in confirming effectiveness.

7.13   Realkalization

In equations (3.1) and (3.2) we saw how carbon dioxide reacts with water to form carbonic acid which then reacts with calcium hydroxide to form calcium carbonate. This removes the hydroxyl ions from solutions and the pH drops so that the passive layer is no longer maintained and corrosion can be initiated.

The cathodic reaction (7.2) showed that by applying electrons to the steel we can generate new hydroxyl ions at the steel surface. This regenerates the alkalinity and pushes the pH back up to around 12.

The anodes used are the same sprayed cellulose or tank anode systems developed to apply chloride removal. In addition to generating hydroxyl ions, the developers claim that by using a sodium carbonate electrolyte they make the treatment more resistant to further carbonation. The patent claims that sodium carbonate will move into the concrete under electro-osmotic pressure. A certain amount will then react with further incoming carbon dioxide. The equilibrium is at 12.2% of 1 M sodium carbonate under atmospheric conditions.

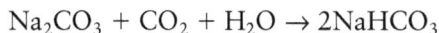

$$Na_2CO_3 + CO_2 + H_2O \rightarrow 2NaHCO_3$$

The method by which the sodium carbonate enters the concrete is contentious with some researchers seeing clear evidence of electro-osmosis Banfill (1994) while others seeing no effect or stating the mechanism to be by diffusion and capillary action Mietz (1998).

In laboratory tests the patent authors have shown that it is very difficult if not impossible for a treated specimen to carbonate again. From the licensor's figures, over 200,000 m^2 of concrete surface area were treated on structures around the world from 1987 to 1999. The treatment is faster than chloride removal only requiring a few days of treatment. The number of treatments has risen significantly since 2000. Standards and guidance documents for realkalisation and ECE have also been published and are discussed in Section 7.13.

7.13.1 Anode types

Anode types are similar to those used for ECE. The sprayed cellulose is used by the owners of the patented system with a steel or coated titanium mesh. The steel is more likely to be used here as the treatment time is shorter and the steel is less likely to be completely consumed.

7.13.2 Electrolytes

As stated earlier sodium carbonate solution is the preferred electrolyte to give long lasting protection against further CO_2 ingress. However, introducing sodium ions can accelerate ASR so in some cases potassium carbonate or plain tap water is used.

7.13.3 Operating conditions

In one case a current density of 0.3–0.5 $A \cdot m^{-2}$ was applied (at 12 V) to 2,000 m^2 of a building in Norway with a treatment time of 3–5 days. In another case 10–22 V was applied to give a current density of 0.4–1.5 $A \cdot m^{-2}$ in 12 days on 300 m^2 of a bridge control tower in Belgium. A further section of 140 m^2 was treated in nine days with a current of 1–2$A \cdot m^{-2}$. All figures are for concrete surface area. The steel to concrete surface ratio was not given. The newly published standard BS EN 14038-1 (2021) states that a minimum of 200 A·h·m^{-2} shall be applied. The previously published draft for development recommended 100 A.h.m^2.

7.13.4 End point determination

This is easy for carbonation. A simple measurement of carbonation depth will show when it has been reduced to zero. However, it has been pointed out that the phenolphthalein indicator turns from clear to pink as the pH rises above about nine. This is still an unpassivated condition. Miller (1989) recommends ensuring that the indicator is bright pink. Universal indicator or an indicator with a colour change closer to 12 may be required to be sure that alkalinity has been fully restored.

7.13.5 Possible effects

As there is a smaller charge density applied compared to ECE the risks of damage are lower than for chloride removal. As mentioned earlier ASR may be a risk if sodium carbonate is used as the electrolyte. Sodium carbonate can also cause short-term efflorescence and the high alkalinity after treatment can attack some coatings. Sodium carbonate will attack oil-based paints, varnishes and natural wood finishes. It should also be noted that carbonated concrete does not promote ASR as the pH has dropped and sodium bicarbonate does not put the pH up to the same level as uncarbonated concrete so the risk of aggravating ASR with realkalization is lower than for ECE.

7.14 Standards and guidance for electrochemical chloride extraction and realkalization

AMPP/NACE had produced and revised state-of-the-art reports on both ECE and realkalization, NACE TR01101 (2018) and NACE TR01104(2020) respectively. It has also produced a standard practice document covering both technologies, NACE SP0107(2021). The European Committee for Standardization (CEN) has produced two standards, originally as drafts for development, technical specifications, now revised and updated as full standards, BS EN 14038-1(2016) and BS EN 14038-2(2021). There is also a Corrosion Prevention Association Technical Note 9 (2018). Much of the original extensive work on ECE was carried out in the Strategic Highway Research Program (SHRP). The report on the laboratory work is SHRP-S-657 (Bennett *et al.*, 1993c) and the field trials are in SHRP-S-669 (Bennett *et al.*, 1993a). A practical implementation guide is published as SHRP-S-

Figure 7.26 An electro-osmosis system consisting of wire anodes and cathodes is shown on a bridge in the UK. There are also rod cathodes in the ground drawing water away.

347 (Bennett, 1993). Recent case histories of ECE treatments are described by Schneck (2006, 2016).

7.15 Electro-osmosis

The phenomenon of electro-osmosis has already been mentioned in connection with electrochemical realkalization (Section 7.8). It is well known that when a porous medium like concrete contains a solution, then an electric current applied between an anode and a cathode will move the water from the anode to the cathode. This leads to drying of anodes for pipelines in soils. The basis of the phenomenon is that when a compound dissolves, water molecules attach themselves to it. This happens more for positively charged metal ions than for negatively charged ions. Therefore more water is carried by the positive ions towards the negative cathode.

It is claimed to be the reason for the transport of the sodium or potassium carbonate from the anode to the cathode in realkalization treatments as discussed in the previous section.

There is therefore proprietary technology for applying this methodology to concrete. The proprietary part is the use of pulses to reduce the build up of charged ions at the electrodes which would increase the electrical resistance. The system works by the installation of cathodes in areas where the water can be discharged and anodes in areas where water must be removed.

The technology is described by McInerney *et al.* (2002, 2004). Figure 7.24 shows a trial system installed in the United Kingdom with horizontal wire anodes and cathodes across the leaf piers and cathodes in the ground to draw water away. The aim of the system was to reduce the relative humidity in the concrete to a level which would not sustain corrosion below say 60% RH (see Section 4.12.4 and Lambert, 1997).

The main problem that the author is aware of with such systems is that the anode is heavily stressed if there is water ingress or run down. Very high current densities can occur on the anodes. This can lead to them failing. It is therefore likely that anodes are in need of further development before the technology can be considered fully developed for external applications where random wetting events cannot be avoided.

In addition, the pulsing process requires sophisticated electronics which must prove durable over many years. They have been found to be less reliable than transformer rectifier systems providing straight DC for impressed current cathodic protection.

Although electro-osmosis systems for drying building materials (e.g. concrete or masonry block basements) are well known technology, its application in reinforced concrete has not progressed from initial trials and installations in the early 2000s. This technology has not been pursued on reinforced concrete since the trial installations in the late 1990s and early 2000s.

7.16   Comparison of techniques

It is possible to summarize the advantages and limitations of the techniques so that they can be compared. This is done more thoroughly in Chapter 8 where Figures 8.1, 8.2 and 8.3 give a breakdown of how to select repairs but the headings below summarize the issues with respect to the electrochemical techniques discussed in this chapter.

7.16.1   Advantages of all electrochemical techniques

1   They treat a large area of a structure, not just patch up immediately corroding areas.
2   They do not give rise to 'incipient anode' problems.
3   Concrete repair to damaged areas is cheaper and easier than for conventional repairs.

7.16.2   Disadvantages of all electrochemical techniques

1   They require specialist knowledge.
2   They must be used with great care and careful selection on pre-stressed structures, coated reinforcement, those with ASR, epoxy injections,

those with poor electrical continuity of the reinforcement, or with 'rogue' metallic inclusions.

7.16.3 Impressed current cathodic protection

1   Advantages: should last 20–100 years or even more with proper maintenance. Proven technology (20 years + and 20–50 year life on anodes), good specifications and standards both for system installation and operation and for embedded and conductive coating anodes (Section 7.4.2)
2   Disadvantages: requires permanent power supply and requires regular maintenance and monitoring and constant power supply.

7.16.4 Galvanic cathodic protection

1   Advantages: very low maintenance, no power supply, no electronics, no wiring. Installation is simplified with no concerns about short circuits from anode to reinforcement. Can be applied to prestressed structures.
2   Disadvantages: no control to ensure that protection is maintained, limited anode life (say 15 ± 5 years). Anodes can be large and intrusive.

7.16.5 Electrochemical chloride extraction

1   Advantages: treatment is completed in six to eight weeks with no further maintenance. Temporary power supply can be used, no mains required.
2   Disadvantages: Limited expertise in practical application. Higher charge density can cause more problems than cathodic protection. Lifetime of treatment not well defined yet. Limited number of practitioners.

7.16.6 Realkalization

1   Advantages: treatment completed in two to four weeks with no further maintenance. Temporary power supply can be used, no mains required. European standard published.
2   Disadvantages: Higher charge density can cause more problems than cathodic protection. Lifetime of treatment not well defined yet. Limited number of practitioners.

7.16.7 Costs

It is difficult to give definite cost information as this varies from job to job and country to country. Summaries of costs for cathodic protection are given in Broomfield (2018) for the United Kingdom and in Bennett *et al.* (1993b) for the USA. These techniques are generally only specified if they offer a cost saving to the owner of the structure over its lifetime

(Unwin and Hall, 1993). Currently in the United Kingdom costs of ECE and realkalization exceed those of impressed current cathodic protection. A broader comparison of the costs of a wider range of repair and treatment methods for the UK can be found at: http://projects.bre.co.uk/ rebarcorrosioncost/. The cost model gives 2003 costs. These will require uprating for inflation but should remain comparable and will be discussed in Chapter 8.

References

AASHTO-AGC-ARTBA (1993). Guide Specification for Cathodic Protection of Concrete Bridge Decks, Task Force 29, Federal Highway Administration, Washington, DC.

American Welding Society (2002). Specification for Thermal Spraying Zinc Anodes on Reinforced Concrete. AWS/ANSI Standard. AWS C2.20/C2.20M: 2018.

Armstrong, K. and Grantham, M.G. (1996). Desalination of Victoria Pier, Jersey. Construction Repair; Jul 1996 Aug 31.

Atkins, C. (2010). Anode System for Harbor Jetties is Adapted to Protect Bridge Deck, Materials Performance, November 2010 NACE International Vol. 49, No. 11.

Baldo, P., Mason, A., Tettamenti, M. and Reding, J. (1991). Cathodic Protection of Bridge Viaduct in the Presence of Prestressing Steel: An Italian Case History, Paper 119, Corrosion 91, NACE International, Houston, TX.

Banfill, P.F.G. (1994). Features of the Mechanism of Re-Alkalisation and Desalination Treatments for Reinforced Concrete. Corrosion and Corrosion Protection of Steel in Concrete, Sheffield, UK.

Bartholomew, J., Bennett, J., Turk, T., Hartt, W.H., Lankard, D.R, Sagues, A.A. and Savinell, R. (1993). Control Criteria and Materials Performance Studies for Cathodic Protection of Reinforced Concrete, SHRP-S-670 Strategic Highway Research Program, National Research Council, Washington, DC.

Bennett, J., Kuan, F.F. and Schue, T.J. (1993a). Electrochemical Chloride Removal and Protection of Concrete Bridge Components: Field Trials, Strategic Highway Research Program, SHRP-5-669, National Research Council, Washington, DC.

Bennett, J., Schue, T.J., Clear, K.C., Lankard, D.L., Hartt, W.H. and Swiat, W.J. (1993c). Electrochemical Chloride Removal and Protection of Concrete Bridge Components: Laboratory Studies, Strategic Highway Research Program Report, SHRP-5-657, National Research Council, Washington, DC.

Bennett, J.E. (1993). Cathodic Protection of Reinforced Concrete Bridge Elements: A State-of-the-Art Report, Strategic Highway Research Program, SHRP-S-337, National Research Council, Washington, DC.

Bennett, J.E. and Broomfield, J.P. (1997). 'Analysis of Studies on Cathodic Protection Criteria for Steel in Concrete'. *Materials Performance*, 36(12): 16–21.

Bennett, J.E. and Schue, T.J. (1993). Chloride Removal Implementation Guide, Strategic Highway Research Program, SHRP-S-347, National Research Council, Washington, DC.

Bennett, J.E. and Schue, T.J. (1998). Cathodic Protection Field Trials on Prestressed Concrete Components, Final Report. FHWA Report. FHWA-RD-97-153.

Bennett, J.E. and Turk, T. (1994). Technical Alert: Criteria for Cathodic Protection of Reinforced Concrete Bridge Elements, Strategic Highway Research Program, SHRP-S-359, National Research Council, Washington, DC.

Bennett, J.E., Bartholomew, J.J., Bushman, J.B., Clear, K.C., Kamp, R.N. and Swiat, W.J. (1993b). Cathodic Protection of Concrete Bridges: A Manual of Practice, Strategic Highway Research Program, SHRP-S-372, National Research Council, Washington, DC.

Bennett, J.E., Bushman, J., Noyce, P. and Costa, J. (2000). Field Application of Performance Enhancing Chemicals to Metallized Zinc Anodes. Corrosion 2000. Paper No. 790.

BRE Digest 330 (1999). Parts 1–4, Alkali-Silica Reaction in Concrete, CRC Ltd, London.

Broomfield, J.P. (1995). Field Measurements of the Corrosion Rate of Steel in Concrete Using a Microprocessor Controlled Guard Ring for Signal Confinement. In N.S. Berke, E. Escalante, C. Nmai and D. Whiting (eds), Techniques to Assess the Corrosion Activity of Steel Reinforced Concrete Structures, American Society of Testing and Materials, STP 1276, Philadelphia, PA.

Broomfield, J.P. (March, 2004). A Case History of Cathodic Protection of a Highway Structure in the UK. Proc. Corrosion 2004. Paper 04344. NACE International Houston, TX.

Broomfield, J.P. (2018). Technical Note 12 Budget Cost and Anode Performance Information for Impressed Current Cathodic Protection of Reinforced Concrete Highway Bridges, Corrosion Prevention Association, Bordon, UK.

Broomfield, J.P. (2018a). Technical Note 19 Acceptable Electrical Resistivities of Concrete Repairs for Cathodic Protection Systems, Corrosion Prevention Association, Bordon, UK.

Broomfield, J.P. (2020). An Overview of Cathodic Protection Criteria for Steel in Atmospherically Exposed Concrete, Corrosion Engineering, Science and Technology 55:4, 303–310 in the UK Proceedings of NACE Corrosion 2004, Paper 04344 AMPP, Houston TX.

Broomfield, J.P. (2021). 'A Historical Review of Impressed Current Cathodic Protection of Steel in Concrete'. *Construction Materials*, 1(1): 1–21.

Broomfield, J.P. and Tinnea, J.S. (1992). Cathodic Protection of Reinforced Concrete Bridge Components, Strategic Highway Research Program, SHRPC/VWP-92-l. 618, National Research Council, Washington, DC.

Broomfield, J.P., Langford, P.E. and McAnoy, R. (1987). Cathodic Protection of Reinforced Concrete: Its Application to Buildings and Marine Structures, Corrosion of Metals in Concrete, Proceedings of Corrosion/87, NACE, Houston, TX, pp 222–235.

Broomfield, J.P. and Wyatt, B.S. (2018). Cathodic Protection of Steel in Concrete, the International Perspective. CPA Technical Note 3 Corrosion Prevention Association, Bordon, Hants, UK.

Brueckner, R., Atkins, C., Foster, A., Merola, R. and Lambert, P. (2012). Maintenance of Transport Structures Using Electrochemical Solutions. Concrete Solutions. Grantham, Mechtcherine & Schneck (eds), 2012 Taylor & Francis Group, London, ISBN 978-0-415-61622-5.

BS EN ISO 12696 (2022). Cathodic Protection of Steel in Concrete, British Standards Institute, London.

BS EN 13509 (2003). Cathodic Protection Measurement Techniques, British Standards Institute, London.

BS EN 14038-1 (2016). Electrochemical Realkalization and Chloride Extraction Treatments for Reinforced Concrete – Part 1: Realkalization, British Standards Institute, London.

BS EN 14038-2 (2016). Electrochemical Realkalization and Chloride Extraction Treatments for Reinforced Concrete – Part 2: Chloride Extraction, British Standards Institute, London.

Buenfeld, N.R. and Broomfield, J.P. (1994). Effect of Chloride Removal on Rebar Bond Strength and Concrete Properties. In Swamy, R.N. (ed.), Corrosion and Corrosion Protection of Steel in Concrete, Sheffield Academic Press, Sheffield, UK.

Came, A. (2018). Surface Preparation of Concrete Prior to Installation of Anode Materials. Technical Note 13, Corrosion Prevention Association, Bordon, UK.

Cáseres, L., Sagüés, A.A., Kranc, S.C. and Weyers, R.E. (2006). 'In Situ Leaching Method for Determination of Chloride in Pore Water'. *Cement and Concrete Research*, 36: 492–503.

CD 370 (2020). Cathodic Protection for Use in Reinforced Concrete Highway Structures. Highways England, London.

Chess, P.M. and Broomfield, J.P. (2014) (Eds) Cathodic Protection of Steel in Concrete and Masonry, 2nd Edition, CRC Press, London.

Clear, K.C., Hartt, W.H., McIntyre, J. and Seung, K.L. (1995). Performance of Epoxy-Coated Reinforcing Steel in Highway Bridges, NCHRP Report 370, National Cooperative Research Program, Transportation Research Board, National Research Council, Washington, DC.

CPA Monograph 10 (2018). Stray Current, Corrosion Prevention Association, Bordon Hants.

Geoghegan, M.P., Das, S.C. and Broomfield, J.P. (1985). Conductive Coatings in Relation to Cathodic Protection, Transport Research Laboratory Report No. P8512100 TRL, Crowthorne, Berkshire, UK.

Glass, G., Taylor, J., Roberts, A. and Davison, N. (2003, Mar.). The Protective Effects of Electrochemical Treatment in Reinforced Concrete. NACE Corrosion 2003. Paper No. 03291 NACE International, Houston, TX.

Glass, G.K., Hassanein, A.M. and Buenfeld, N.R. (2000). 'CP Criteria for Reinforced Concrete in Marine Exposure Zones'. *Journal of Materials in Civil Engineering*, 12(2): 164–171.

Hassanien, A.M.G., Glass, G.K., Buenfeld (2002), N.R. Protection Current Distribution in Reinforced Concrete Cathodic Protection Systems. *Cem. Concrete Composition*, 24: 159–167.

Hime, W.G. (1994). 'Chloride-Caused Corrosion of Steel in Concrete: A New Historical Perspective'. *Concrete International*, 15(5): 56–61.

ISO (2018). Accelerated Life Test Method of Mixed Metal Oxide Anodes for Cathodic Protection – Part 1 Application in Concrete, ISO 19097-1, ISO Geneva Switzerland.

Kessler, R.J., Powers, R.G. and Lasa, L.R. (1995). Update on Sacrificial Anode Cathodic Protection of Steel Reinforced Concrete Structures in Seawater. Corrosion 95, Paper 516, NACE International, Houston, TX.

Kessler, R.J., Powers, R.G. and Lasa, I.R. (2002). An Update on the Long Term Use of Cathodic Protection of Steel Reinforced Concrete Marine Structures. NACE Corrosion 2002. Paper No. 02254.

Lambert, P. (1997). 'Controlling Moisture'. *Construction Repair*, 11(2): 29–32.

Manning, D.G. (1995). 'Corrosion Performance of Epoxy Coated Reinforcing Steel: North American experience'. *Construction and Building Materials*, 10(5): 433.

McInerney, M.K., Cooper, S.C., Hock, V.F. and Morefield, S.W. (2004). Measurements of Water and Ion Transport in Concrete via Electro-Osmosis. Proc. Corrosion 2004. Paper 04351. NACE.

Mears, R.B. and Brown, R.H. (1938). 'A Theory of Cathodic Protection'. *Trans. Electrochemical Society*, 74: 519–531.

Michael, M., Sean, M., Sondra, C., Phillip, M., Charles, W., Matthew, B., Jonathan, T. and Vincent, F.H. (2002). Electro-Osmotic Pulse (EOP) Technology for Control of Water Seepage in Concrete Structures. ERDC/CERL TR-02.

Mietz, J. (1998). Electrochemical Rehabilitation Methods for Reinforced Concrete Structures – A State of the Art Report. 24 ed. London: European Federation of Corrosion. Publication/Institute of Materials. ISBN: ISSN 1345–5116.

Miller, J.B. (1989). Note on the Use of Phenolphthalein as a pH-indicator for Carbonated Concretes. Norwegian Concrete Technologies Paper No. 7.

Morgan, T.D.B. (1990). Some Comments on Reinforcement Corrosion in Stagnant Saline Environments. In Page, C.L., Treadaway K.W.J. and Bamforth P.B. (eds), Corrosion of Reinforcement in Concrete, Elsevier Applied Science, London for the Society of Chemical Industry, pp 29–38.

Morrison, G.L., Virmani, Y.P., Stratton, F.W. and Gilliland, W.J. (1976). Chloride Removal and Monomer Impregnation of Bridge Deck Concrete be Electro-osmosis. Kansas DoT Report. FHWA-Ks-RD. 74–1.

NACE 11100 (2018). Reference Electrodes for Atmospherically Exposed Reinforced Concrete Structures, Association of Materials Protection and Performance, Houston, TX.

NACE International Publication 01105 Sacrificial Cathodic Protection of Reinforced Concrete Elements, A State-of-the-Art Report. 2005; Item No. 24224.

NACE RP0169 Control of External Corrosion on Underground or Submerged Metallic Piping Systems.

NACE RP0100-2000 Cathodic Protection of Prestressed Concrete Cylinder Pipelines. NACE Standard Recommended Practice. 2000 March.

NACE RP0290-2000. Impressed Current Cathodic Protection of Reinforcing Steel in Atmospherically Exposed Concrete Structures. NACE Standard Recommended Practice. 2000 Jun.

NACE SP0107 (2021). Electrochemical Realkalization and Chloride Extraction for Reinforced Concrete, Association of Materials Protection and Performance, Houston, TX.

NACE Standard (2001). TM0294-2004.Testing of Embeddable Impressed Current Anodes for Use in Cathodic Protection of Atmospherically Exposed Steel-Reinforced Concrete.

NACE Standard Test Method (2005). TM105-2005(21247). Test Procedures for Organic-Based Conductive Coating Anodes for Use on Concrete Structures.

NACE TR01101 (2018). Electrochemical Chloride Extraction from Steel-Reinforced Concrete – A State of the Art Report, Association for Materials Protection and Performance, Houston, TX.

NACE TR01102 (2018). State of the Art Report: Criteria for Cathodic Protection of Prestressed Concrete Structures, Association of Materials Protection and Performance, Houston, TX.

NACE TR01104 (2020). Electrochemical Realkalization of Steel-Reinforced Concrete – A State-of-the-Art Report, Association of Materials Protection and Performance, Houston, TX.

Pianca, F., Schell, H., Lai, D. and Raven, R. (2003). The Ministry of Transportation of Ontario's Experience with Electrochemical Chloride Extraction (ECE) 1989–2002. NACE Corrosion 2003. Paper No. 3293.

Polder et al (2014). 'Service Life and Life Cycle Modelling of Cathodic Protection for Concrete Structures'. *Cement and Concrete Composites*, 47: 69–74.

Pourbaix, M. (1973). Lectures on Electrochemical Corrosion, Plenum Press, New York.

Sagüés, A.A. (1994). Corrosion of Epoxy Coated Rebar in Florida Bridges, Final Report to Florida DOT, WPI No. 0510603, College of Engineering, University of South Florida, Tampa, FL.

Schneck, U. (2006). Experiences and Conclusions from 5 years of Chloride Extraction Applications. Concrete Solutions, Proceedings of the 2nd International Conference, St Malo, France.

Schneck, U. (2016). Focused, Nondestructive Repair of Tunnel Walls, Concrete Solutions, Proceedings of the 16th International Conference, Thessaloniki, Greece.

Schneck, U. (2019). Instant Performance Verification after Electrochemical Chloride Extraction by Enhanced Corrosion Testing. Proc. 7th Concrete Solutions Conference, Cluj Napoca, Matec Conferences, 2019.

Schneck, U. (2021). Private correspondence.

Sergi, G. and Page, C.L. (1992). The Effects of Cathodic Protection on Alkali-Silica Reaction in Reinforced Concrete, Contractor Research Report 310, Transport Research Laboratory, Crowthorne, Berkshire, UK.

Slater, J.E., Lankard, D.R. and Moreland, P.J. (1976). 'Electrochemical Removal of Chlorides from Concrete Bridge Decks'. *Materials Performance*, 15(11): 21–25.

Society for the Cathodic Protection of Reinforced Concrete (1995). Status Report: The Cathodic Protection of Reinforced Concrete, Society for the Cathodic Protection of Reinforced Concrete, Leighton Buzzard, UK.

Sohanghpurwala, A.A. (2003). Long-Term Effectiveness of Electrochemical Chloride Extraction on Laboratory Specimens and Reinforced Concrete Bridge Components. NACE Corrosion 2003. Paper No. 3293.

Stratfull, R.F. (1974). Experimental Cathodic Protection of a Bridge Deck. Transportation Research Record, 500, Transportation Research Board, Washington, DC.

Unwin, J. and Hall, R.J. (1993). Development of Maintenance Strategies for Elevated Motorway Structures. In M.C. Forde (ed.), Structural Faults and Repair 93, University of Edinburgh, Engineering Technics Press, 1: 23–32.

Virmani, Y.P. and Jones, W.R. (1988). 'Analysis of Concrete Cores from a Cathodically Protected Bridge Deck'. *Public Roads*, 51(4): 123–127.

Chapter 8

Rehabilitation methodology

One of the major issues facing any consultant or owner of a structure suffering from chloride or carbonation-induced corrosion is what form of repair to undertake. As we have seen from the previous sections there are coatings, sealants, membranes and enclosures, specialized patch repair materials, options for total or partial replacement, impressed current and galvanic cathodic protection, electrochemical chloride removal, realkalization and corrosion inhibitors. These can be applied to structures suffering different degrees of corrosion due to chloride attack or carbonation or a combination of these two. Each treatment will have implications for the future maintenance requirements, time to next major intervention and ultimate service life of the structure.

The practical solution for most owners of individual corroding structures will be to take advice from a civil engineering consultant or corrosion specialist who has extensive experience of working on the durability problems of atmospherically exposed concrete structures. Advice may also be sought from materials suppliers and applicators about their own particular systems and a consensus will be reached about the most effective repair to the structure based on local knowledge, experience and availability of materials and systems.

Many owners of large structures and large inventories of structures will have developed in-house knowledge. They will have consulted widely and may develop a systematic approach to corrosion rehabilitation. Bridge management systems, although no longer in their infancy, are still not specifically geared to detailed corrosion assessment or its management, particularly for reinforced concrete bridge elements.

The SHRP program developed a manual and a computer program aimed at providing highway agencies with a way of carrying out life cycle cost analysis for corrosion rehabilitation of reinforced concrete bridges (Purvis *et al.*, 1994). The approach makes many assumptions based on the knowledge and experience of the authors and has been developed further by FHWA but it has found limited acceptance. However, it was probably the first comprehensive systematic attempt to select a rehabili-

DOI: 10.1201/9781003223016-8

tation method based on engineering judgement rather than subjective judgement.

8.1   Technical differences between repair options

Different repair options are suitable for different applications, however, these overlap to such an extent that a clear methodology for making a decision on technical or cost grounds is difficult as each structure and type of structure is different. What is acceptable on a bridge deck is not applicable and acceptable on a substructure or a building. Costs of access vary widely as do the implication of noise, vibration and dust.

Table 8.1 gives indications of the compatibilities of different repair approaches on decks and vertical surfaces based on the different requirements of the different systems. Giving global information is difficult because there are different approaches in the United States, Europe and other regions to concrete repair and the requirements of each structure and each location and microclimate must be assessed individually.

A simplified table of technical issues that may exclude some repair options is also given later. This lists some issues such as the design of the structure and other problems that may eliminate certain options or may severely increase the cost of using them.

A repair methodology is given in the three figures 8.1 to 8.3. They require first an evaluation of the structure using techniques as discussed in Chapter 4, then a choice between the available options. In all cases there is the 'do nothing' option. This says that corrosion is not too severe and the benefits of waiting outweigh the benefits of repairing or rehabilitating now. This option can be attractive for very short times but the structural and safety implications of ongoing corrosion must be fully assessed and understood. Whichever strategy is chosen, its merits and limitations should be examined as done in Chapters 6 and 7. European and US Standards, specifications and guidance documents for the techniques have been given in each chapter.

Chapter 9 will discuss how we can evaluate the current condition, develop a condition index and extrapolate forwards and backwards to predict future deterioration.

8.2   Repair costs

Providing general cost guidance is difficult especially as cost information is difficult to find for a generalized approach. Even on a specific project, comparing costs for different repair options is difficult if not impossible.

This is further complicated in making allowance for inflation effects and life-cycle costing. The normal US practice is to use a factor of about 3% as the normal difference between inflation and interest rates. However, in the past the United Kingdom the Treasury has required a 7%

Table 8.1a Decks – Bridge, car park, jetty

Problem	Overlay	Waterproof	Patching	ICCP	GCP	ECE/realk	Comments
Dead load or clearance problems	Difficult	Difficult (needs overlay)	Ok	Ribbon, probe anodes or soffit anode	Ok	Ok	Overlays add dead weight and cut clearance
Overlay exists	Replace	Replace (with asphalt)	With replacement	Soffit or mesh plus overlay	Ok	Need to remove and replace overlay	Working through overlays is difficult
Polymer injection exists	Ok	Ok	Ok	Not possible	Not Possible	Not possible	Treatment cannot reach reinforcement
No mains electricity	Ok	Ok	Ok	Only possible with alternative power supply	Ok	Ok with temporary power	ICCP requires permanent power
Poor rebar continuity	Ok	Ok	Ok	Could be expensive	Ok – not ideal	Could be expensive	Electrochemical techs require rebar continuity
ASR problems	Watch alkali content of overlay	Ok	Ok	Ok	Ok	Test for problem and use LiBO$_4$	Increase of alkali could exacerbate ASR
Prestressing	Ok	Ok	Ok	Difficult	Ok	Very difficult	ICCP cathodic reactions risk hydrogen embrittlement

Table 8.1b Walls, façades, columns

Problem	Overlay/enclose	Waterproof/coat/impregnate	Patching	ICCP	GCP	ECE/realk	Reason
Historic façade	Unlikely to be acceptable	Unlikely to be acceptable could use silane or silicate	Difficult cosmetic matching short and long term	Suitable anodes are ribbon, probe or placed on inside	Unlikely, but could place inside	Ok	Issues of aesthetics and removal of fabric
Noise, vibration, dust and so on			Could use hydrojetting to minimize	Good as repair is minimized	Good as repair is minimized	Good as repair is minimized	
Marine	Unlikely to be successful	Unlikely to be successful	OK (short life without additional protection)	Ok	Ok	Unlikely to work (sea water distributes current away from rebar)	Aggressive marine exposure

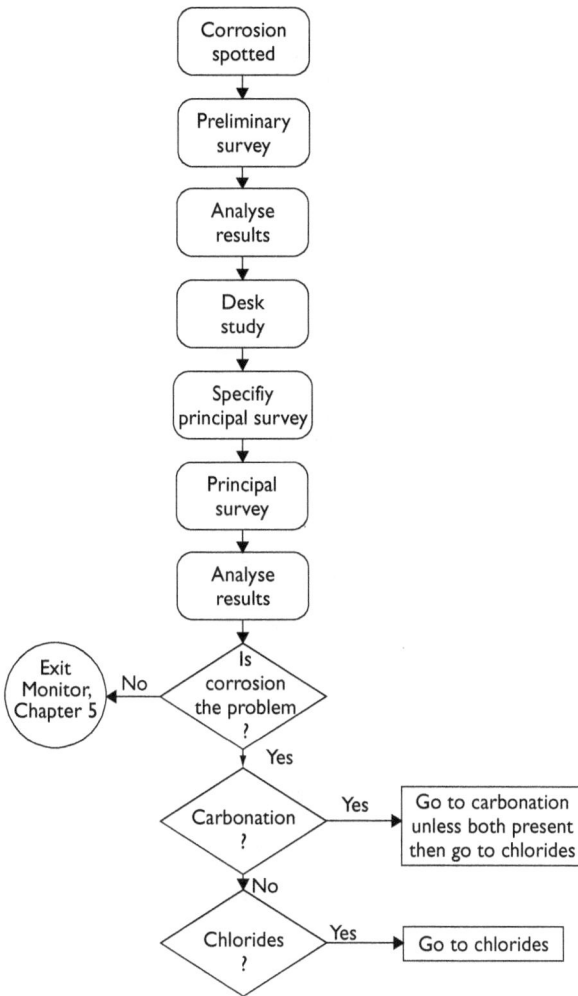

Figure 8.1 Flow diagram for evaluation of a corroding structure prior to repair selection.

discount rate as the difference between inflation and the cost of a commercial loan (Unwin and Hall, 1993). This had a highly distorting effect on all life-cycle costing for UK highway structures. The effect of the high UK rate of inflation at time of writing has yet to be seen.

The most coherent set of comparative data was developed by the SHRP research on repair methods (Gannon *et al.*, 1993). A series of field rehabilitation projects on US bridges was investigated and the data presented in a series of graphs and polynomial equations. In the United Kingdom a web-based model for evaluation of structures, selection of repair options

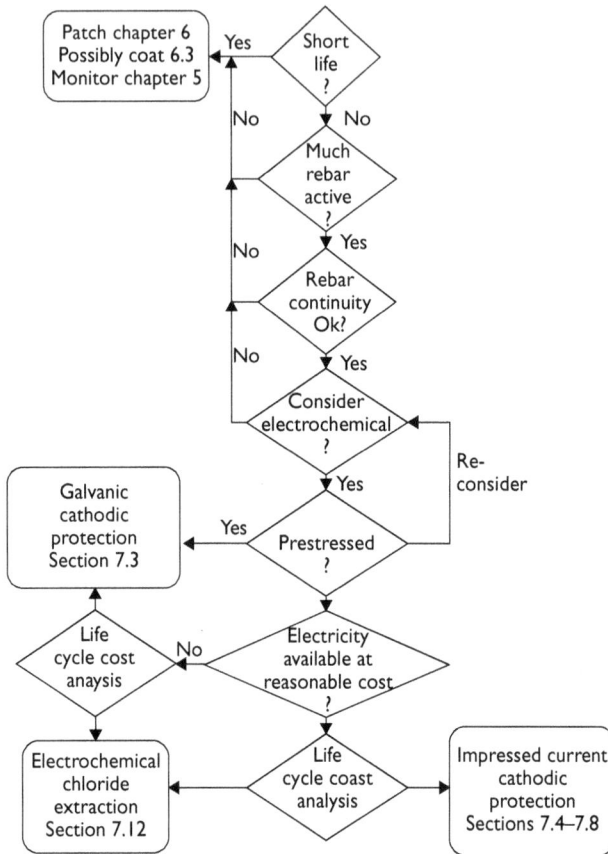

Figure 8.2 Flow diagram for selecting treatment for chloride-induced corrosion.

and comparative life-cycle costing. The model is available for use at http://projects.bre.co.uk/rebarcorrosioncost/. This includes a comprehensive set of cost tables for galvanic and impressed current cathodic protection, electrochemical chloride extraction, electrochemical realkalization, surface coatings, patch repair and permanent corrosion monitoring at 2000 prices and includes a simple engine for generating life cycle costs from calculated repair costs and repair life times.

8.3   Carbonation options

For carbonation the usual options are patch repairing usually with a suitable anticarbonation coating applied afterwards (Section 6.3.1) and

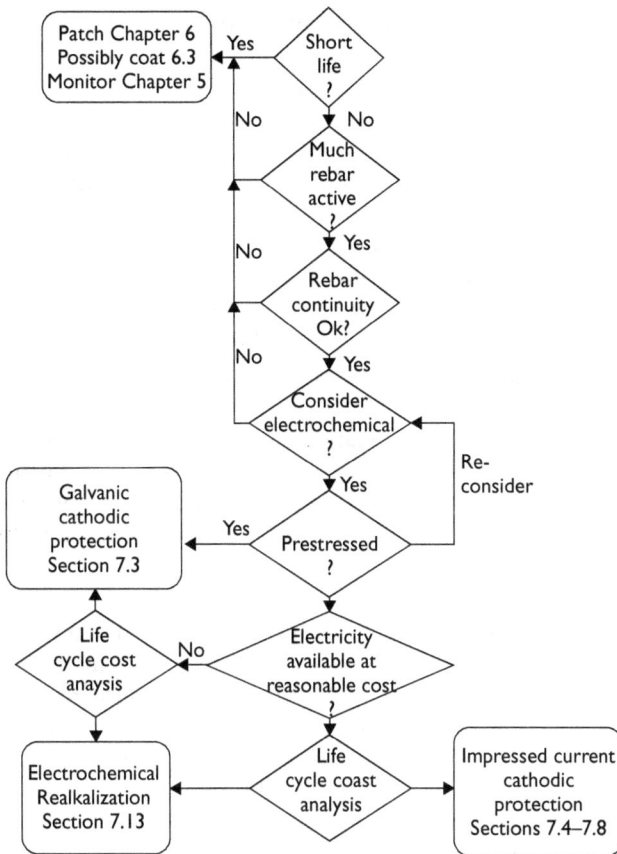

Figure 8.3 Flow diagram for selecting treatment for carbonation-induced corrosion.

realkalization (Section 7.12). Realkalization is very difficult in the presence of prestressing. The latest evidence is that ASR is not a significant problem. Realkalization (and all electrochemical techniques) becomes less cost effective if there are large number of unconnected rebars. Realkalization is most cost effective for larger surface areas and structures requiring long times to the next major intervention.

A third option is now available in the form of corrosion inhibitors. These can be applied to the concrete surface, to the broken out area to be patch repaired and in the patch repair material. However, there is still considerable ambiguity in the literature as to their effectiveness as discussed in Section 6.6.

The rest of this section will discuss the options of patching and coating, realkalization and inhibitor application for carbonation repairs.

8.3.1   Patching and coating

Patching and coating is used for repairing carbonation and chloride-induced corrosion damage. They are generally more successful for carbonation as was shown in a recent study by Seneviratne (2000) and Sergi (2000) slowing the rate of corrosion as well as the rate of advance of the carbonation front.

The effect of moisture reduction was studied by Tuutti (1982) and is discussed in Section 4.12.4. He found a 'critical degree of saturation' for carbonated concrete, where the corrosion rate jumped by almost an order of magnitude. This was at 85% RH. The corrosion rate then rose by two more orders of magnitude to peak at about 95% RH. This implies that if the relative humidity stays below 85% (or 80% for a margin of safety), then the corrosion rate will be low (0.001–0.001 $\mu A\ cm^{-2}$). For chloride ingress, the curves are less clear but suggest a significant level of corrosion above 60% RH, peaking nearer 90% RH. This explains why anticarbonation coatings can control carbonation-induced corrosion rather than just slow the advancing carbonation front. This may be due to the fact that only a modest reduction in relative humidity is needed to slow the corrosion rate initiated by carbonation, to about 80% RH, as opposed to the need to reduce it to below 60% for chloride-induced corrosion.

Patching is far more effective for carbonation than for chlorides. This is because restoring the alkalinity is all that is necessary to stop corrosion in carbonated concrete and an anticarbonation coating will then slow carbon dioxide ingress as well as reducing the moisture content. The discussion in Section 6.2 on patching shows how the 'incipient anode effect' can actually accelerate corrosion around a patch. This phenomenon is usually observed in chloride conditions rather than carbonation, which is another reason why patching is more effective in carbonated rather than chloride contaminated structures. Contaminated or carbonated concrete must be removed from all around the steel. In some cases temporary structural support may be needed during the repair process.

An accurate bill of quantities is needed for accurate price quotations and estimates. It is therefore important to define the areas and volume of repair accurately. One of the biggest problems encountered in carrying out patch repairs is that the amount of patching is underestimated. This is usually for two reasons, the inaccuracy of the 'sounding' technique for delimiting areas of delamination and the growth of those areas in the time between surveying and the repair contractor carrying out the work. Chapter 9 discusses how the estimate can be revised with time, especially if there are excessive delays.

8.3.2   Why choose realkalization?

As stated earlier, patching and applying an anticarbonation coating is much more effective than patching and coating for chloride attack. It was the only repair option for carbonated structures until the early 1990s. The problems of continued outbreaks of corrosion that had led to the development of cathodic

protection for chloride contaminated structures had not manifested themselves for carbonated structures. However, in the late 1980s realkalization was
developed and it is a recognized alternative method of rehabilitating structures
where carbonation is a significant and recognized problem (Section 7.12).

The extent to which carbonation has reached the rebar and the
requirements for patch repairing to restore alkalinity will determine whether
realkalization is preferred either because it is more economic or in other
cases to avoid the noise, dust and vibration required for extensive patch
repairing. As the patches can be quite shallow (see Figure 6.1), temporary
structural support may be avoided as the concrete behind the steel is not
removed. The technique is becoming popular in Europe. In North America
very little attention is paid to carbonation. Whether this is a difference in
climate, concrete or perception is not clear. The amount of attention dedi-
cated to the chloride problem on highway bridges may have led to the minimizing of the perception of the problem of carbonation in North America.

Figure 8.4 shows the growth in realkalization treatment from the
development of the patented process in 1987–1999. The figures for chloride removal have been included but were a very small proportion of the
totals in the period shown. Since then it has been harder to acquire information on treatment rates but ECE now seems more popular than realkalization due to improved anticarbonation coatings and patch repair
materials. The company CITec in Germany states that it has treated 40
structures and 4,700 sq m up to 2021 with ECE. The CPA Technical
Note 9 on realkalization states that over 180,000m² of reinforced concrete surface was treated with realkalization between 1986 and 2000.

The main questions to be asked when considering this option are:

• How long will it last? Is that time compatible with the service life
  requirements of the structure?

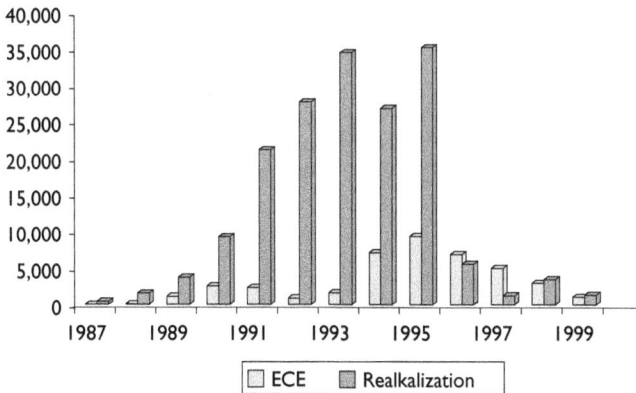

Figure 8.4 Annual figures for the treatment of reinforced concrete by ECE and realkalization
in square metres. Figures Courtesy Fosroc (UK) Ltd.

- Is repeat treatment feasible and acceptable if carbonation resumes at the end of the service life of the treatment (say after 10 years)?
- Are there any problems with my structure that could interact with the realkalization process?

The main requirements for realkalization are:

- electrical continuity of the steel;
- reasonable level and uniformity of conductivity of the concrete when wet;
- no metallic (electronic) short circuits to the anode are present;
- no risk of causing or accelerating alkali-silica reactivity;
- no risk of hydrogen embrittlement of prestressing steel;
- availability of electrical power for the treatment period;
- no electrically insulating layers between the surface and the reinforcing steel.

These requirements can be checked as an extension of the condition surveyor as a specific feasibility survey/study. If there are problems with continuity or shorts or high resistance patch repairs then these can often be dealt with but the cost should be analysed to ensure that the process is still cost effective.

The problem of hydrogen embrittlement and prestressing steel has been discussed in Section 7.9 for cathodic protection, and similar issues apply for realkalization. The realkalization process applies 20–40 V DC between the anode and the steel. It must therefore send the steel potential well beyond the level needed for hydrogen evolution.

Great care must therefore be exercised when applying realkalization to structures containing prestressing steel under load. Very careful monitoring must be carried out to ensure that the steel is completely shielded from the risk of hydrogen evolution or the steel must be carefully tested to ensure that it has no susceptibility to embrittlement from nascent hydrogen. These issues are discussed in Miller (1994).

Due to the problems with prestressing steel, realkalization would probably not be used as the alternative of patch repairing and coating would provide more cost-effective solutions. In any case, it would be unusual for carbonation to take place in prestressed structures due to the high cement contents and high cover in such elements. If carbonation had reached prestressing steel in such structures there would probably be concerns about the strength of the concrete.

8.4 Chloride options

If chlorides are the cause of corrosion then patching, overlaying, encasement, cathodic protection and electrochemical chloride extraction are the most likely options. As discussed previously all techniques have different merits and limitations. This section explores them and gives some comparisons.

8.4.1 *Patching and sealing*

Some degree of patching will probably be required whatever other repair is required. The merits and limitations of patch repairs and coatings are different for chloride contaminated structures compared to carbonated structures. If chloride ingress is limited and effective patch repairs can be undertaken, removing chlorides to below the corrosion threshold and controlling moisture and chloride ingress then they can give cost-effective life extension. However, while coatings have been shown to control corrosion in carbonated structures, this has not been observed for structures where corrosion is due to chlorides Seneviratne (2000) and Sergi (2000).

Since the pores in concrete contain significant amounts of water and air, sealing it to stop corrosion is unlikely to be effective once chlorides have penetrated the concrete. There is also a hygroscopic effect where chlorides in concrete form salts that absorb moisture. Figure 4.19 shows the curve of corrosion rate vs. chloride content, the curve is less clear than 4.20 for carbonation but suggests a significant level of corrosion above 60% RH, peaking nearer 90% RH. This suggests that coatings will not control chloride-induced corrosion in most environments where chloride ingress is a problem as ambient relative humidities are above 60% (London UK humidity ranges from 56% to 87% with an average of 72% RH).

In the case of chlorides, patch repairs are only effective if chloride ingress is local and the chlorides can all be removed. Just patching up damaged areas is a short-term palliative not a long-term rehabilitation. This is due to the incipient anode effect discussed in Section 6.2.1. It was found from corrosion monitoring of bridges with sealers and pore blockers or even just sealed joints after years of chloride ingress that there was an increase in resistivity and a move towards more passive half cell potentials for beams where sheltered under the deck.

There are two exceptions to the requirement for patch repairing as part of the rehabilitation process:

1 Chlorides have been detected and a preventive treatment such as cathodic protection is to be applied before delamination starts.
2 The level of cracking and spalling is acceptable and cathodic protection or chloride removal is to be applied as a 'holding' treatment to prevent further deterioration.

The amount of patch repairing necessary and its underestimation are discussed in the previous section and above under carbonation repairs.

8.4.2 Why choose impressed current cathodic protection?

This technique has been applied to reinforced concrete since the 1970s with the first trials reported over 60 years ago (Stratfull, 1959). Anode systems are still developing while many commercially available systems have track records of more than 40 years. No deleterious effects have been found over that period apart for one anode system that failed after about 5 years and is no longer marketed for concrete applications. Cathodic protection has been applied to concrete bridges in the United States since 1973 (Stratfull, 1985) and in the United Kingdom to buildings and other structures since 1985 (Broomfield et al., 1987). It has been stated by the US Department of Transportation that cathodic protection is the only method of stopping corrosion across the whole structure regardless of chloride content FHWA-RD-01-096 (2001).

There are now many national, international and proprietary specifications for impressed current cathodic protection for steel in concrete. It can only be applied with great care if there is prestressing in the structure. Careful investigation is recommended in the presence of ASR. There are additional installation costs if there are large numbers of electrically unconnected reinforcing bars.

The main advantages of impressed current cathodic protection for chloride attack are as follows:

- It prevents corrosion damage across the whole of the area where the cathodic protection system is applied.
- Corrosion should not recur throughout the life of the system.
- It has a long (>60 year) track record of successful application to bridges and other reinforced concrete structures above ground, with hundreds of structures and thousands of square metres treated.
- There are established, non-proprietary specifications for its application.
- There are anode systems suitable for most applications.
- Competitive tendering is routine in most countries.
- Life-cycle cost analysis has shown it to be cost effective for a wide range of structures and conditions.

The following checklist should be considered for cathodic protection and is similar to that given above for realkalization:

- Is there electrical continuity of the steel?
- Is there reasonable level and uniformity of conductivity of the concrete?
- Is there a risk of causing or accelerating alkali-silica reactivity?
- Is there a risk of hydrogen embrittlement of prestressing steel?
- Is electrical power available for the life of the structure?
- Is there a management system in place for monitoring and controlling the system?

Continuity can be remedied prior to installation if necessary.

8.4.3 Why choose galvanic cathodic protection?

Galvanic cathodic protection has its own advantages and disadvantages relative to impressed current cathodic protection and the other electro-chemical and conventional rehabilitation techniques. The different anode systems also have their own merits and limitations.

The main advantages of galvanic cathodic protection are:

* It does not require power
* It does not require monitoring, control or maintenance
* It can be applied to prestressing and in the presence of ASR
* There is no problem of short circuiting anodes
* It can be installed as an 'add on' to a patch repair.

Its limitations are:

* It cannot be guaranteed to achieve the control criteria of BS EN ISO 12696 or NACE SP0290 which ensure corrosion is controlled
* It can be expensive to install some anode types
* It can dry out in some locations and microclimates
* Anodes have unknown and uncontrolled lifetimes.

8.4.4 Why choose chloride removal?

Cathodic protection has been described as the only solution to chloride-induced corrosion that can stop (or effectively stop) the corrosion process. The problems with cathodic protection are its requirement for a permanent power supply, regular monitoring and maintenance.

Similarly, chloride removal can stop corrosion across the whole struc-ture and has the advantage that, like a patch repair, it is a one off treat-ment. A generator can be brought in for the duration of the treatment so mains power is not essential. There is no long-term maintenance need but the system does treat the whole structure.

The disadvantage is its unknown duration of effectiveness. We cannot remove all the chlorides from the concrete. If we can stop further chlo-ride ingress then the system may be effective for many years (10–20). If chlorides are still impinging on the structure then the time to re-treat-ment will be shorter (or may require a coating or sealant).

Chloride removal cannot be applied to prestressed structures due to the risk of hydrogen embrittlement. The use of lithium-based electrolytes suggests that ASR can be controlled. As stated earlier for impressed cur-rent cathodic protection, there must be electrical continuity within the reinforcement network for any of the electrochemical techniques to be applied. We do not know how long the treatment process will last but a range of 5–20 years is likely, depending upon conditions.

The main advantages of chloride removal are as follows:

- It prevents corrosion damage across the whole of the area treated
- There are a number of different anode systems suitable for different applications
- Life cycle cost analysis has shown it to be cost effective for a wide range of structures and conditions
- It does not require a permanent power supply. A diesel generator or other temporary source can be used
- It does not require regular monitoring (although checks on reinitiation of corrosion may be needed on a yearly or two yearly basis).

The following checklist should be considered for chloride removal and is similar to that given above for realkalization and cathodic protection:

- Electrical continuity of the steel
- Reasonable level and uniformity of conductivity of the concrete
- No metallic (electronic) short circuits to the surface are present
- No risk of causing or accelerating alkali-silica reactivity
- No risk of hydrogen embrittlement of prestressing steel.

8.4.5 Other chloride repair options

The other chloride repair options include:

- Inhibitors (see Section 6.6)
- Concrete Overlays (used on North American highway bridges according to specifications developed by highway agencies)
- Sealers (more appropriate before corrosion is initiated)
- Waterproofing membranes (used before corrosion has initiated)
- Coatings, barriers and deflection systems.

The last three systems are best applied well before corrosion has initiated. As the chloride profile within the concrete approximates to a square root relationship, there will be considerable reserves of chloride in the cover concrete to push the concentration at rebar level above the corrosion threshold even if the supply is completely cut off. It may be advisable to calculate the redistribution of chlorides to see if and when redistribution will lead to depassivation.

Total enclosure in an 'indoor' environment will control chloride-induced corrosion as long as the relative humidity is maintained below 60% (Figure 4.19). The exact threshold requires further research and the relationship between RH and chloride content.

8.5 Standards and guidance for selection of repairs

We have seen that the selection of a suitable rehabilitation can be based on technical considerations and cost (preferably whole life costing rather than just initial installation cost). The technically unacceptable can be excluded and a shortlist of suitable rehabilitations can be drawn up. Life cycle cost analysis techniques have been used to calculate the optimum time and the optimum repair on bridges but are based on a number of assumptions and estimates, including cost estimates for different repair strategies. A direct comparison of quotations for a given structure is probably the best present state of the art although it is important to accurately define the corrosion conditions so that accurate bills of quantities can be drawn up.

A major European standard for repair methods, selection specification and installation has now been completed with the publication of EN1504 parts 1–10. This includes a large suite of test standards for certification of the materials. These documents are designed to meet the harmonized products directive and are principally concerned with the materials used in non-electrochemical concrete repair, their testing prior to use and the site application. Table 8.2 lists the standards. From the point of view of selection and application of repairs parts 9 and 10 are key parts of this standard. A commentary on BS EN 1504 parts 1 to 10 and the associated test methods is given by Raupach and Buttner (2014) and in Concrete Society (2009).

In part 9, structural repairs and corrosion repairs are covered separately. The corrosion methodology is summarized in Table 2 *Principles and Methods Relating to Reinforcement Corrosion*. This is reproduced as Table 8.3.

A systematic method of using EN1504 would be as follows:

- EN1504-9 summarizes the evaluation process covering both structural and reinforcement corrosion issues. The corrosion evaluation is covered in Chapter 4 of this book and a flow diagram is given in Figure 8.1.
- EN1504-9 summarizes the options. These are covered in Chapters 5, 6 and 7 and flow diagrams Figures 8.2 and 8.3 summarize the approach. EN1504-9 discusses:

  - 'Do Nothing' option, with or without monitoring (see Chapter 5)
  - Structural issues (beyond the scope of this book)

Table 8.2 Products and systems for the protection and repair of concrete structures –
definitions, requirements, quality control and evaluation of conformity

Part	Name	Comment
1	Definitions	First published 2005
2	Surface protection systems for concrete	Lists the following applications for surface protection: protection against ingress, moisture control, physical resistance/surface improvement, resistance to chemicals and increasing resistivity by limiting moisture content. Lists treatments as impregnation, hydrophobic impregnation or coating and lists performance requirements to achieve the above
3	Structural and non-structural repair	Covers mortars and concrete applied by hand, recasting, spraying, overlaying and patching. Requirements covered include thermal compatibility, elastic modulus, skid resistance, coefficient of thermal expansion and capillary absorption. Materials are divided into structural and non-structural based on compressive strength adhesion and elastic modulus
4	Structural bonding	Covers plate bonding, us of precast units and adhesives for bonding fresh concrete to existing concrete
5	Concrete injection	Materials for filing cracks, voids, interstices for strengthening, waterproofing or protection against ingress
6	Anchoring of reinforcing steel bar	Covers strengthening structures by casting new parts with bars anchored in original structure
7	Reinforcement corrosion prevention	This part covers primers and protective coatings for the reinforcing bars applied as part of a patch repair
8	Quality control and evaluation of conformity	Sets out the requirements for materials producers to comply with the standard and achieve CE marking
9	General principles for the use of products and systems	See following table and text for full description
10	Site application of products and systems and quality control of the works	Covers both the requirements for the specification of material works and for their application on site. Includes surface preparation, compatibility of material and substrate and between materials, application conditions and achievement of specified properties

- Prevention or reduction of further deterioration (see Section 6.3)
- Improvement strengthening or refurbishment of all or part of the structure (see Chapters 6 and 7)
- Full or partial reconstruction (generally beyond the scope of this book but see Section 6.1 on concrete removal).

Table 8.3 BS EN 1504-9 Products and systems for the protection and repair of concrete structures – definitions, requirements, quality control and evaluation of conformity

Principle No.	Principle and definition	Examples of methods based on the principle
7 [RP]	Preserving or restoring passivity Creating chemical conditions in which the surface of the reinforcement is maintained in or is returned to a passive condition	7.1 Increase cover to reinforcement with additional cementitious mortar or concrete 7.2 Replacing contaminated or carbonated concrete 7.3 Electrochemical realkalization of carbonated concrete 7.4 Realkalization of carbonated concrete by diffusion 7.5 Electrochemical chloride extraction
8 [IR]	Increase resistivity Increasing the electrical resistivity of the concrete	8.1 Hydrophobic impregnation 8.2 Impregnation 8.3 Coating
9 [CC]	Cathodic control creating conditions in which potentially cathodic areas of reinforcement are unable to drive an anodic reaction	9.1 Limiting the oxygen content (at the cathode) by saturation or surface coating
10 [CP]	Cathodic protection	10.1 Applying an electrical potential
11 [CA]	Control of anodic areas	11.1 Active coating of the reinforcement 11.2 Barrier coating of the reinforcement 11.3 Applying inhibitors in or to the concrete

Source: Part 9 General principles for the use of products and systems Table 1: Principles and Methods for protection and repair of concrete structures.

- EN1504-9 then lists a series of factors to be considered when choosing a rehabilitation option
    - intended use of structure
    - performance characteristics (e.g. water tightness, fire resistance)
    - repair life time
    - opportunities for additional repair and monitoring
    - life-cycle costing throughout the residual life of the structure
    - performance and properties of the existing substrate
    - final appearance.
- There are additional issues such as
    - health and safety
    - environmental impact

    – appropriateness of the selected option(s)
    – conformity with the standard.

- The concrete rehabilitation materials for non-electrochemical techniques can then be listed and the required properties (bond strength, shrinkage, elasticity, strength, etc.) selected from EN1504 parts 2 to 7.
- The production of a specification and application conditions and methods are given in EN1504-10. The most useful information being in the informative annexes.
- Finally EN1504-9 requires the compilation of a record of the works and an inspection and maintenance procedure.
- Use Figure 8.1 and Chapter 4 determine the cause of deterioration.
- Use Figures 8.2 or 8.3 to short list suitable rehabilitation techniques (see also Table 8.1).
- Use Sections 8.3 and 8.4 EN1504-9 Table 1 (Table 8.3) to narrow down options.
- If selected options are electrochemical and are not covered by EN1504 then go to the relevant European Standard or ISO as given in Chapter 7 (or use a AMPP/NACE national or government agency standard or specification if more relevant or if a suitable EN document does not exist).
- If the repair option falls within EN1504 then use EN1504-10 to determine the relevant properties and application methods of the materials to be applied.

8.6 Training

From the content of this book, it can be seen that anyone carrying out assessment, repairs and designs for corrosion control systems should be adequately trained and have relevant experience. The author has seen cases of cathodic protection designs that are totally unsuitable for reinforced concrete and repairs that failed within a few years of application. The designers may have been competent in other areas but did not have the specialist expertise and sufficient information to deal with the problems in the structures in question.

There have been major developments in the training and certification of those designing, applying and maintaining corrosion control systems in general and specifically for corrosion of steel in concrete. For many years the NACE corrosion courses in CP and coatings led the way internationally. Other national corrosion bodies, such as the Institute of Corrosion in the UK, had similar courses. Most of these were general courses or were aimed at specific industries such as the petrochemical industry. However, the development of the European standard, now BS EN ISO 15257 (2017), gives specific training for the design, installation, monitoring and maintenance of cathodic protection and related technology for a range of applications, with a specific set of courses for CP of steel in concrete. There are five levels of certification, paraphrased below:

- Level 1 – CP Tester – shall be competent to collect CP performance data of simple CP systems and perform other basic CP tasks in accordance with technical instructions and shall understand the fundamentals of the measurements that they are required to undertake.
- Level 2 – CP Technician – shall be competent to undertake a range of CP measurement, inspection and supervisory activities in accordance with technical instructions and procedures and shall have knowledge of the fundamentals of electricity, corrosion, coatings, CP and measurement techniques, safety issues and applicable standards.
- Level 3 – CP Senior Technician – shall provide guidance for persons at Level 1 and Level 2. They shall be competent to prepare technical instructions for all CP persons of lower-level competence and assess all data collected from these tasks.
- Level 4 – CP Specialist – shall have detailed knowledge of corrosion theory, principles of electricity, CP design, installation, commissioning, testing and performance evaluation, including systems affected by interfering conditions. They shall have a general familiarity with CP in all application sectors. They shall be competent to design CP systems.
- Level 5 – Cathodic Protection Expert – Level 5 CP persons shall have advanced the state of the art of CP by scientific work and peer-reviewed publications and shall have made a marked and original contribution to the science or practice of corrosion control by CP.

There has been discussion in the UK by the Structural Concrete Alliance (SCA) of a 'Level zero' qualification for site operatives installing anodes and wiring and other CP components into concrete. There is ongoing discussion of a suitable qualification being developed as a National Vocational Qualification (NVQ). At the moment the author is not aware of any National Corrosion body offering a process for achieving Level 5.

As well as attending a training course and passing the examination, the applicant must also show experience to the level required and produce evidence of continued professional development (professional development hours) and ongoing projects in the field to maintain certification.

The certification schemes based on the standard have been developed in a number of European countries, including France, Germany and the UK. The UK institute of Corrosion has also trained and provided certification to other nationals. Although NACE was involved in the development of the ISO standard, they have not adopted it, and at the time of writing they do not have a specific course on steel in concrete. However, the steel in concrete content of the AMPP/NACE CP courses has significantly increased.

When assessing and repairing a structure it is important to have adequately trained and experienced professionals designing, reviewing and installing the systems. AMPP and ISO based certification should be requested of the team members or they should provide a resumé showing equivalent levels of knowledge and experience.

References

Broomfield, J.P., Langford, P.E. and McAnoy, R. (1987). Cathodic Protection for Reinforced Concrete: Its Application to Buildings and Marine Structures NACE. Corrosion of Metals in Concrete. San Francisco. Houston, TX: NACE; 1987, pp. 222–235.

BS EN ISO 12696 (2022). *Cathodic Protection of Steel in Concrete*, British Standards Institute, London.

BS EN ISO 15257 (2017). Cathodic Protection – Competence Levels of Cathodic Protection Persons – Basis for Certification Scheme. British Standards Institute, London.

CEN (2021). Electrochemical Realkalization and Chloride Extraction Treatments for Reinforced Concrete – Part 1: Realkalization. BS EN 14038-1:2021.

Concrete Society (2009). Repair of Concrete Structures with Reference to BS EN 1504, Technical Report 69, The Concrete Society, Camberley, UK.

EN1504-1 – Products and Systems for the Protection and Repair of Concrete Structures – Definitions, Requirements, Quality Control and Evaluation of Conformity: Part 1 Surface Protection Systems for Concrete.

EN1504-10 – Products and Systems for the Protection and Repair of Concrete Structures – Definitions, Requirements, Quality Control and Evaluation of Conformity: Part 10 Site Application of Products and Systems and Quality Ccontrol of the Works.

EN1504-2 – Products and Systems for the Protection and Repair of Concrete Structures – Definitions, Requirements, Quality Control and Evaluation of Conformity: Part 2 Surface Protection Systems for Concrete EN1504-3 – Products and Systems for the Protection and Repair of Concrete Structures – Definitions, Requirements, Quality Control and Evaluation of Conformity: Part 3 Structural and Non-structural Repair.

EN1504-4 – Products and Systems for the Protection and Repair of Concrete Structures – Definitions, Requirements, Quality Control and Evaluation of Conformity: Part 4 Structural Bonding.

EN1504-5 – Products and Systems for the Protection and Repair of Concrete Structures – Definitions, Requirements, Quality Control and Evaluation of Conformity: Part 5 Concrete Injection.

EN1504-6 – Products and Systems for the Protection and Repair of Concrete Structures – Definitions, Requirements, Quality Control and Evaluation of Conformity: Part 6 Anchoring of Reinforcing Steel Bar.

EN1504-7 – Products and Systems for the Protection and Repair of Concrete Structures – Definitions, Requirements, Quality Control and Evaluation of Conformity: Part 7 Reinforcement Corrosion Prevention.

EN1504-8 – Products and Systems for the Protection and Repair of Concrete Structures – Definitions, Requirements, Quality Control and Evaluation of Conformity: Part 8 Quality Control and Evaluation of Conformity.

EN1504-9 – Products and Systems for the Protection and Repair of Concrete Structures – Definitions, Requirements, Quality Control and Evaluation of Conformity: Part 9 General Principles for the Use of Products and Systems.

FHWA-RD-01-096 Long Term Effectiveness of Cathodic Protection Systems on Highway Structures. Federal Highway Administration, McLean VA USA.

Gannon E.J., Cady, P.D. and Weyers, R.E. (1993). Concrete Bridge Protection and Rehabilitation: Chemical and Physical Techniques, Price and Cost Information, Strategic Highway Research Program, SHRP-S-664 National Research Council, Washington, DC.

Miller, J.B. (1994). Structural Aspects of High Powered Electrochemical Treatment of Reinforced Concrete. In Swamy R.N. (ed.), Corrosion and Corrosion Protection of Steel in Concrete, 2: 1499–1511, Sheffield Academic Press, Sheffield, UK.

NACE RP0290. Impressed Current Cathodic Protection of Reinforcing Steel in Atmospherically Exposed Concrete Structures. NACE Standard Recommended Practice. 2000 Jun.

Purvis, R.L. Babaei, K., Clear, K.C. and Markow, M.J. (1999). Life Cycle Lost Analysis for Protection and Rehabilitation of Concrete Bridges Relative to Reinforcement Corrosion. SHRP Published Report. SHRP-S-377. National Research Council, Wastington, DC.

Raupach, M. and Buttner, T. (2014). Concrete Repair to EN 1504, CRC Press, Boca Raton, FL.

Seneviratne, A.M.G., Sergi, G. and Page, C.L. (2000). 'Performance Characteristics of Surface Coatings Applied to Concrete for Control of Reinforcement Corrosion'. *Construction and Building Materialism*, 14: 55–9.

Sergi, G., Seneviratne, A.M.G., Maleki, M.T., Sadegzadeh, M. and Page, C.L. (2000). 'Control of Reinforcement Corrosion by Surface Treatment of Concrete'. *Proc. Instn. Civ. Engrs., Structures & Buildings*, 140(1): 85–100.

Stratfull, R.F. (1959). 'Progress Report on Inhibiting the Corrosion of Steel in a Reinforced Concrete Bridge'. *Corrosion*, 15(6): 331t–334t.

Stratfull, R.F. (1985). Eleven Years of Success for Coke Breeze C.P. Pavement. Cathodic Protection of Reinforced Concrete Bridge Decks. pp. 66–81.

Tuutti, K. (1982). Corrosion of Steel in Concrete. Swedish Cement & Concrete Research Institute, Stockholm.

Unwin, J. and Hall R.J. (1993). Development of Maintenance Strategies for Elevated Motorway Structures. Structural Faults and Repair 93. (Ed. Forde, M.C.), 1: 23–33

Chapter 9

Modelling and calculating corrosion, deterioration and life cycle costing of reinforced concrete structures

Corrosion of steel in concrete can be modelled as a three-stage process. The first, usually called the initiation stage, is the diffusion of CO_2 or chlorides to the steel to cause depassivation. The second stage is the 'activation stage'. This is a less clear cut phenomenon as more of the rebar network starts to corrode, and the rust products are formed. The third stage is deterioration as cracking and spalling occurs. Eventually a situation is reached which is defined as limit state or the end of the functional life and rehabilitation must take place by that point. The process is illustrated in Figure 9.1 using an exponential function and a simple linear fit.

9.1 Activation time T_0 carbonation-induced corrosion

Carbonation and chloride-induced corrosion are both generally considered to be diffusion-based phenomena. The rate of progress of carbonation is given by the equation:

$$d = At^n \tag{9.1}$$

where d = carbonation depth in mm, A is a coefficient, t is time in years and n is an exponent, usually $= 1/2$. This is a simple derivation of the diffusion law that says the rate of diffusion is proportional to the thickness being diffused through:

$$\frac{dx}{dt} = \frac{1}{kx}$$

where x is the thickness and t is time. Therefore $dt = kx \cdot dx$ by integration $t = kx^2 + k_0$.

DOI: 10.1201/9781003223016-9

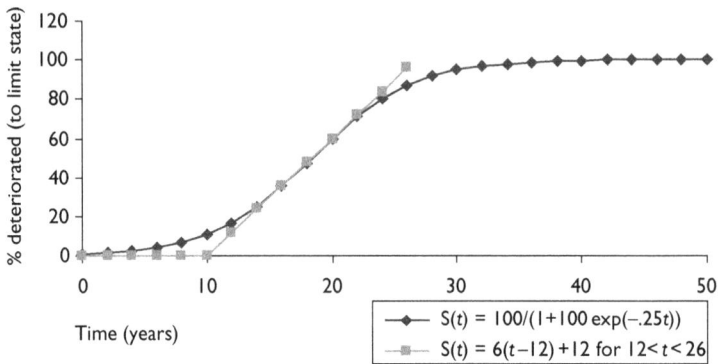

Figure 9.1 Deterioration curves and condition indices.

In Scandinavia, the constant of integration k_o is used to reflect any initial maltreatment that can lead to initial carbonation. For instance an average quality concrete that was deshuttered too early and may have had slight surface freezing can give $k_0 = 10$ mm. A properly treated concrete has $k_o \leq 1$ mm (Miller, 1995).

Assuming that there is no carbonation at time zero $x = AT^{1/2}$. This is based on the assumptions of steady state, that is, constant CO_2 content at the surface and constant, with uniform conditions within the concrete.

Equation (9.1) is a very simple equation to use if we can derive a value for the constant. However, there are deviations from the parabolic relationship. These include changing conditions such as humidity and temperature to change the coefficient, giving apparent deviation from the $t^{1/2}$ behaviour. Also the concrete quality changes with depth so the difficulty of diffusion will be variable. This leads to the generalized equation:

$$X = At^n \quad \text{where } 0 \leq n \leq 1$$

A number of empirical calculations have been used to derive values of A and n based on such variables as exposure conditions (indoors and outdoors, sheltered, unsheltered), 28-day strength and water cement ratio. A wider range of empirically derived equations is given in Table 3.1. These cover different exposure conditions, curing and concrete properties. The easiest solution for a given structure is to take some measurements of carbonation depth, assume $n = 1/2$ and calculate A. This can be used to predict the rate of progression of the carbonation front. The time taken to reach the steel can then be estimated and the rate of depassivation calculated.

9.1.1   Determination of carbonation rates

As stated in the previous section, the easiest way to predict the carbonation rate of an existing structure is to measure the carbonation depths at representative locations and use equation (9.1) to calculate the coefficient A, assuming $n = 1/2$ or a chosen value. A simple spreadsheet programme for doing this can be found at http://projects.bre.co.uk/rebarcorrosioncost/ and is described in Trend 2000 *et al.* (2001).

However, when there is no structure and life cycle modelling is required, two approaches are possible, one is described in Parrott and Hong (1991) and Parrott (1994a,b). This calculates the corrosion rate based on air permeability of the concrete or of a trial mix of the concrete.

Parrott analysed the literature (Parrott, 1987), see Section 3.1.1, and suggested that the carbonation depth D at time t is given by:

$$D = ak^{0.4} t^n - C^{0.5}$$

where k is air permeability (in units of 10^{-16} m^2), C is the calcium oxide content in the hydrated cement matrix for the cover concrete, $a = 64$. k can be calculated from the value at 60% relative humidity by the equation:

$$k = mk_{60}$$

where $m = 1.6 - 0.00115r - 0.0001475r^2$ or $m = 1.0$ if $r < 60, n = 0.5$ for indoor exposure but decreases under wetter conditions to:

$$n = 0.02536 + 0.01785r - 0.0001623r^2$$

To measure the concrete cover depth required to prevent carbonation from reaching the steel it is therefore necessary to measure the air permeability and the relative humidity, and then calculate D. This can be done with a proprietary apparatus developed by Parrott and available commercially. Gas and air permeability measurements are discussed in Kropp and Hilsdorf (1995).

An alternative approach is to be found in Bamforth (2004). This model will be discussed in more detail in Chapter 10. At this point we can say that by analysing the literature Bamforth found that the key parameters affecting the carbonation rate were:

1   the buffering capacity of the concrete (relative to a benchmark Portland cement concrete)
2   the C_3A content of the concrete
3   equivalent period of wet curing
4   mean relative humidity of the environment.

The relevant parameters for polynomial curves relating a specific concrete mix to a 'standard' concrete mix were derived for a range of blended and composite cements and developed into a model for predicting carbonation rates for new construction. A spreadsheet programme is provided on CD ROM supplied with published copies of Bamforth (2004).

9.2   Chloride ingress rates (initiation)

For chloride ingress life is more complex as we are dealing with atmospherically exposed structures that are exposed to variable chloride contents. Also, we are not dealing with a front that moves through the concrete but a chloride profile that builds up in the concrete.

The usual form of the diffusion equation used is Fick's second law:

$$\frac{d[Cl^-]}{dt} = D_c \frac{d^2[Cl^-]}{dx^2}$$

where $[Cl^-]$ is the chloride concentration at depth 'x' at time 't' and D_c is the diffusion coefficient (usually of the order 10^{-10} m^2/sec).

The exact solution to this equation is:

$$\frac{C_{max} - C_d}{C_{max} - C_{min}} = erf \frac{(x)}{(\{4D_c t\}^{1/2})} = u \tag{9.2}$$

where u is the concentration ratio for each depth d below the maximum concentration C_{max} at or near the surface, C_{min} is the background level of cast in chloride (assumed to be evenly distributed) and erf( ) is the error function of the value of the argument in brackets.

In atmospherically exposed concrete there is no easy initial or surface concentration as the chlorides at the surface can vary from zero to 100% depending upon wetting, drying, evaporation, wash off, etc. It is therefore common practice to discard the first 5–10 mm of a chloride profile sample and take the next increment, around the 10 mm depth, as a constant initial, pseudo surface concentration. If this is done then diffusion calculations must use the depth from the sampling depth, not from the surface.

9.2.1   The parabolic approximation

A simplified method of calculating the initiation time for chloride attack is to look at the progress of the 'chloride threshold' through the concrete. By taking samples with depth it is possible to fit a parabolic curve to the chloride concentration (or more simply to fit a straight line to a plot of depth vs. the square root of chloride concentration) and to find the depth at which the concentration is 0.4% chloride by weight of cement (or 1 lb per cubic yard or whatever threshold is chosen). Its rate of progress through the concrete can then be predicted using the simple diffusion equation (9.1). Estimates of the time to reach the rebar can then be made (or the more complex equation (9.2) can be used). However, any background concentration of chloride must be allowed for.

However, an additional complication comes from the observation that the diffusion constant changes with time. It is therefore important to only take measurements on 'mature' concrete as the diffusion rate may be higher in the first few years.

A more sophisticated approach to using the parabolic approximation was developed by Poulsen (1990) who pointed out that within defined mathematical limits the error function expression for the diffusion coefficient can be approximated to the simple parabolic function. He used the following expression for Fick's second law:

$$C_{(x, t)} = C_i + (C_s - C_i)\mathrm{erfc}[x/(4tD_0)^{1/2}]$$

where $C_{(x, t)}$ is the chloride concentration at time t and depth x (the profile), C_i is the initial chloride concentration, C_s is the surface chloride concentration, $\mathrm{erfc}(z)$ is the error function complement and D_0 is the diffusion constant.

By using the approximation $\mathrm{erf}(z) = (1 - z/\sqrt{3})^2$

$$C_{(x, t)} = C_i + (C_s - C_i)[1 - x/(12tD_0)^{1/2}]^2 \tag{9.3}$$

for $0 \leq x \leq (12tD_0)^{1/2}$ y, and for $x > (12tD_0)$ y, then $C_{(x, t)} = 0$. The equation (9.3) is in the form $y = mx + c$. It can therefore be solved by simple graphical techniques.

If a series of incremental drillings is taken, then a plot of square roots of the concentration change $(C_s - C_i)$ vs. distance into the concrete can be made. A straight line can be fitted, ignoring any surface effects due to washing out of chlorides or concentration due to evaporation of water leaving excessively high levels of chloride behind. The error limits are a very useful check to exclude extreme values and ensure that the values are on a parabolic curve approximating to a diffusion curve.

The straight line graph can be used to calculate the diffusion coefficient and the rate of movement of any required chloride threshold in mm yr^{-1}. The process is recommended for use by Germann Instruments with their RCT profile grinder (see Section 4.10 and Figure 4.10). It is fully described with examples in their instruction and maintenance manual (Germann Instruments, 1994).

9.2.2 Andrade on resistivity vs diffusion coeff

In a more recent paper, Torres Acosta et al (2019) found good correlation between resistivity measurements, Rapid Chloride Permeability Tests ASTM C1202 (2019), and the apparent chloride diffusion coefficient. The resistivity measurements were taken from compressive strength cylinders for new construction and cores from the field. This provides a quicker and more efficient method of finding the chloride diffusion coefficient for new construction and existing structures.

9.2.3 Sampling variability for chlorides

Inevitably an inhomogeneous material like concrete will show variation in the chloride content unless very large samples are taken due to the variation in the ratio of paste to aggregate. There will also be some variation due to the local ability of the matrix to resist chlorides. Further, in many atmospherically exposed structures there will be very large differences in chloride content due to the differences in exposure. For instance, on a bridge substructure there will be areas of water rundown that are exposed to very high levels of chlorides while adjacent areas are comparatively unaffected. Chloride laden water may pond at some sites (e.g. on the top of beams, especially if they are horizontal). At the bottom of the beam the water may evaporate leaving the chloride behind.

This can make it very difficult to sample consistently or identify typical environments. It is made even more difficult if intervention has started, for example, by applying gutters under the joints, building up concave cross heads, etc. where water had previously ponded. These early interventions can prevent easy identification of areas susceptible to high or low chloride ingress.

The minimum recommended sample size for chloride analysis is 25 g of sample. CBDG (2002) recommends that if drilled samples are being taken then a 25 mm diameter drill bit is used to collect samples at 25 mm increments. Concrete Society (2004) suggests using two nominal 20 mm diameter holes 30–40 mm apart and combining the samples. This is the procedure recommended in the European standard on chloride testing of hardened concrete, BS EN 14629. Sampling variability and the taking and grinding of core samples is discussed in Section 4.10 and illustrated in Figure 3.2(b).

9.2.4 Mechanisms other than diffusion

It is important to recognize that diffusion is not the only transport mechanism for chlorides in concrete, particularly in the first few millimetres of cover. There may be several mechanisms moving the chlorides including capillary action and absorption as well as diffusion. Rapid initial absorption occurs when chloride laden water hits very dry concrete. In many circumstances these will only affect the first few millimetres of concrete. If so then the expedient of ignoring the first few millimetres of drillings and then calculating diffusion profiles will work. If the cover is low, the concrete cycles between very dry and wet or the concrete quality is low then the alternative transport mechanisms may overwhelm diffusion, at least to rebar depth.

This rapid initial absorption of chlorides may help to explain the over estimate of the diffusion rate often made when predicting future chloride contamination rates from data collected in the first few years of service.

Both carbonation and chloride diffusion rates are functions of temperature. As seasonal and daily variations are rapid compared with diffusion of chlorides or carbon dioxide, these are assumed to average out. The environments in northerly and southerly latitudes change in temperature, relative humidity and exposure condition so separating out temperature effects is difficult. A full discussion of all transport mechanisms is given in Kropp and Hilsdorf (1995).

9.3   Rate of depassivation (activation)

We have treated the problem as one dimensional so far, considering the time to depassivation at one particular location. Carbonation depths, chloride profiles and rebar depths are not uniform so the spatial distribution of depassivation or initiation must be included in the calculation unless the ranges are small or the time from depassivation to damage is large. We know that all the concrete cover will not spall off at once so there must be a distribution of depassivation times and of time from depassivation to spalling. We must have realistic estimates of the time from the first spall to end of functional service life.

By looking at the distribution of diffusion coefficients and the cover distribution it should be possible to calculate T_0 for the first 1, 10, 20%, etc. of the structure. It may also be important to differentiate between different locations due to variations in exposure. This will include moving up a column from the sea level, areas of salt water run off on substructures, zones facing salt spray, etc.

9.4   Activation time T_1

The diffusion models work reasonably well for predicting the initiation time. The chloride profile and the carbonation depth can be measured in the field or from cores in the laboratory. However, it is far more difficult to look at the next step in our model. Corrosion rate measurements are now being taken in the field with linear polarization instruments and empirical estimates have been made with different instruments for the time to spalling.

The following broad criteria for corrosion have been developed from field and laboratory investigations with the sensor controlled guard ring device (Feliu et al., 1995):

Passive condition $I_{corr} < 0.1\ \mu A/cm^2$
Low to moderate corrosion I_{corr} 0.1 to 0.5 $\mu A/cm^2$
Moderate to high corrosion I_{corr} 0.5 to 1 $\mu A/cm^2$
High corrosion rate $I_{corr} > 1\ \mu A/cm^2$.

The device without sensor control has the following recommended interpretation (Clear, 1989):

No corrosion expected $I_{corr} < 0.2\ \mu A/cm^2$
Corrosion possible in 10 to 15 years I_{corr} 0.2 to 1.0 $\mu A/cm^2$
Corrosion expected in 2 to 10 years I_{corr} 1.0 to 10 $\mu A/cm^2$
Corrosion expected in 2 years or less $I_{corr} > 10\ \mu A/cm^2$.

The relative sensitivities of the two instruments are discussed in the section on corrosion rate measurement. These measurements are affected by temperature and R.H., so the conditions of measurement will affect the interpretation of the limits defined earlier. The measurements themselves should be considered accurate to within a factor of two (Feilu *et al.*, 1995).

For carbonation it seems that the rate of corrosion falls rapidly as the relative humidity in the pores drops below 75%, and rises rapidly to an R.H. of 95% (Tuutti, 1982). There is also approximately a factor of 5–10 reduction in corrosion rate with 10°C reduction in temperature (see Section 4.12). The calculation of corrosion rates is therefore critically dependent upon having measurements at different times of year and under different (representative) conditions to derive a realistic typical or average corrosion rate.

The corrosion currents given earlier can be directly equated to section loss by Faraday's law of electrochemical equivalence, where $1\ \mu A \cdot cm^{-2} = 11.5\ \mu m$ section loss/year.

Various efforts have been made to estimate the amount of corrosion that will cause spalling. It has been shown that cracking is induced by less than 0.1 mm of steel section loss, but in some cases far less that 0.1 mm have been needed. This is a function of the way that the oxide is distributed (i.e. how efficiently it stresses the concrete), the ability of the concrete to accommodate the stress (by creep, plastic or elastic deformation) and the geometry of rebar distribution that may encourage crack propagation by concentrating, stresses, etc., for example, in a closely spaced series of bars near the surface, or at a corner where there is less confinement of the concrete to restrain cracking.

What we refer to as 'Rust' is a complex mixture of oxides, hydroxides and hydrated oxides of iron having a volume ranging from twice to about six times that of the iron consumed to produce it (see Figure 2.2). This assumes 100% density of the product. Any porosity will increase the volume further.

If we assume a volume increase of three on average (or four allowing for the 1 : 1 replacement of the consumed steel) then the corrosion rates above translate as follows:

Table 9.1 Equations published in the past from researchers regarding time to concrete cracking due to corrosion of embedded reinforcement. The equations have been modified from the original papers for uniformity in presentation (Aligizaki, 2006)

Researchers	Equation derived	Comments
Bazant (1979)	$t_2 = \rho_{cor} \dfrac{d_0 \cdot \Delta t}{s \cdot MR}$, $\rho_{cor} = \left[\dfrac{1}{\rho_{ox}} - \left(\dfrac{0.583}{\rho_{st}} \right) \right]^{-1} \dfrac{\pi}{2}$	Theoretical model
Ravindrarajah and Ong (1987)	$\Delta M = 0.487(2c + d_0) \cdot \dfrac{d_0}{c} + 21.7$	Impressed current, one quality of concrete, iron loss is not Q_{cr}
Wang and Zhao (1993)	$\dfrac{\Delta t}{\Delta d_{cr}/2} = 0.33 \cdot \left(\dfrac{d_0}{c} \right)^{0.565} f_c^{1.436}$	Laboratory experiments and finite element analysis
Morinaga et al. (1994)	$Q_{cr} = 0.602 \left(1 + \dfrac{2c}{d_0} \right)^{0.85} d_0$	Impressed current
Rodriguez et al. (1994)	$\Delta d_{cr} = \alpha_1 \cdot 10^{-13} \cdot \left[83.8 + 7.4 \cdot \left(\dfrac{c}{d} \right) - 22.6 \cdot f_{t,sp} \right]$ where $\Delta d_{cr} = d_0 - d$	Impressed current, one concrete quality, chlorides in the initial mix
Torres-Acosta and Sagüés (1998)	$\Delta d_{cr} = 14.4 \times 10^{-4} \cdot \left(\dfrac{c}{L} + 1 \right)^{1.48} \left(\dfrac{c}{d_0} \right)^{0.87} d_{aggr}$	Experimental

Notes
CR = Corrosion rate
MR = Mass loss rate
Q_{cr} = Amount of corrosion to cause cracking
c = Cover thickness
d_0 = Initial diameter of steel bar
d = Final diameter of steel
d_{aggr} = Size of aggregates
$f_{t,sp}$ = Splitting tensile strength of concrete
f_c = Cube compressive strength of concrete
L = Length of corroding bar
s = Spacing of steel bars
t_2 = Propagation period
ΔM = Mass loss of iron
Δd_{cr} = Reduction of steel diameter due to corrosion to cause concrete cracking
Δt = Increase in bar diameter
α_1 = Coefficient that depends on type of corrosion
ρ_{ox} = Density of the oxide
ρ_{st} = Density of steel.

0.1 μA·cm^{-2} = 1.1 μm/year section loss = 3 μm/year rust growth
1.0 μA·cm^{-2} = 11.5 μm/year section loss = 34 μm/year rust growth
5.0 μA·cm^{-2} = 57.5 μm/year section loss = 173 μm/year rust growth
10.0 μA·cm^{-2} = 115 μm/year section loss = 345 μm/year rust growth.

From the corrosion rate measurements it would appear that about 10 μm section loss or 30 μm rust growth is sufficient to cause cracking.

A number of theoretical and empirical methods for calculating the time to cracking have been derived these are compiled from Table 9.1 taken from Aligizaki (2006). Other authors have used finite element analysis models (Tabatabai and Lee, 2006).

9.5 The Clear/Stratfull empirical calculation

Stratfull developed an empirical equation to determine the time to first distress of reinforced concrete in sea water with a known, constant chloride content. Clear modified this to be used for atmospherically exposed structures.

$$T = \left[\frac{(0.052\, d^{1.22}\, t^{0.21})}{(Z^{0.24} P)} \right]^{0.83} \tag{9.4}$$

where T = time to first cracking (years); d is the depth of cover in mm or depth of cover minus depth of near surface chloride measurement (Z); t is the age at which Z was measured and P is the water/cement ratio. One difficulty with using this model is knowing when it is no longer valid as it is empirically based on US interstate bridge structures. Other variations on this equation exist (Purvis *et al.*, 1992). It is very difficult to compare models although much work is going on in developing them.

9.6 Corrosion without spalling and high corrosion rates

A different situation exists when high levels of water saturation and low levels of oxygen lead to the 'black rust' that is deposited in the concrete without exerting stresses (see Section 2.2). In this case the corrosion rate measurement can be taken as showing a section loss that will eventually lead to failure. However, it is very difficult to extrapolate an instantaneous measurement of corrosion rate to a total section loss measurement. If we can make a series of corrosion rate measurements at different locations and we can come up with a compensation for variations in relative humidity and temperature we can estimate the average corrosion rate. We also have to estimate the original time to corrosion, and assume that the corrosion rate has either been constant or increased in some sensible manner (say linear or logarithmic) to the present condition.

It is also very difficult to measure corrosion rates in these situations as they are usually underwater or under waterlogged conditions such as a failed membrane.

If an investigation reveals pits then one assumption is that the corrosion rate is about five times that measured with an accurate linear polarization device. There is wider discussion of this issue in Section 4.12.4.

9.7   Cracking and spalling rates, condition indexes and end of functional service life

We can predict the initiation time, and the time to cracking and spalling and give a distribution of that time to show how corrosion damage will spread with time and location across our structure. In order to decide at what point it must be repaired we must define an 'end of service life condition'. This is not an easy definition. It is often subjective and will vary from structure to structure. It will also vary from engineer to engineer. In an SHRP study of service life estimates where 60 North American highway engineers were questioned about their bridges it was found that bridge decks had reached the end of their functional service life when 5–14% of the deck was spalled, delaminated or patched with temporary asphalt patches. For substructures it was found that about 4% cracking, spalling and delamination represented end of functional service life (Weyers et al., 1994).

In a separate SHRP study (Purvis et al., 1994) a condition index S was derived:

$$S = \frac{[Cl + 2.5(Delam) + 7.5(Spall)]}{8.5} \tag{9.5}$$

where Cl is percentage of bars with chloride above the corrosion threshold, Delam is the percentage of surface with hidden delaminations and Spall is the percentage of surface with visible deterioration.

No attempt was made to define an end of service life as this was used as part of a life-cycle cost analysis study where repair or rehabilitation was conducted according to the minimum cost criterion. It was suggested that the condition index should never exceed 45%.

Some engineers will define the end of functional service life or the limit state as the time to first spall as this is the point at which remedial action will be undertaken. From a safety point of view this may be relevant but it may not be cost effective depending upon the importance of the appearance of the structure and the feasibility of containing and controlling concrete debris. Obviously a spall on a deck is of less concern than one from the top of a multi-storey building.

The link between cracking and spalling is difficult to define. However, it may be possible to look at the cracking rate and suggest that if 5% of bars have cracks above then this will lead to 1% spalling. The condition index equation (9.5) earlier suggests a 3 : 1 ratio of cracking to spalling. The SHRP researchers suggested that the percentage delamination is four times the spalling percentage for most concrete decks, eight times for decks with ⩾25 mm concrete overlays and for most substructures and 16 times for substructures with ⩾25 mm concrete jackets or shotcrete.

The authors also suggested that chloride contamination (percentage above the corrosion threshold) increases linearly from 0 at condition index 0–100% at condition index 20.

In the SHRP study the condition index was calculated as a function of time by fitting it to a curve of the form:

$$S_t = \frac{100}{[1 + A\exp(-Bt)]} \qquad (9.6)$$

This is an 'S' shaped curve that approximates to the two straight lines for T_0 and T_1 in Figure 9.1. The curve could therefore be calculated if there are two sets of data. These can be derived from two sets of measurements of cracking, spalling and chloride content separated by several years, or taking one set of measurements at the present time t_p and back calculating data for an earlier time t_e. This may be to the time of the first spall or a back calculation of the time to depassivation from the chloride profiles (approximately T_0). We can therefore derive values for A and B. These can be used to project forward the delamination rate and show how costs will escalate if work is deferred or how repair quantities will increase between the survey and the start of patch repair work.

9.8   Summary of methodology to determine service life

When presented with a corroding structure we can determine its condition by measuring the chloride profiles, carbonation depths and cover depths. From this we can calculate diffusion rates of the carbonation front or the chloride threshold and estimate the initiation time T_0. Both an average and a distribution of T_0 values can be derived.

By measuring corrosion rates over a period of time we can estimate the time to cracking and knowing the distribution of corrosion rates, cover depths, etc. a cracking rate can be established by adding the time to cracking to the initiation time. An empirical condition curve can be calculated and the time taken to reach an unacceptable level can be determined.

9.9 Diffusion models proposed in literature

As stated earlier, there are a number of models for corrosion of steel in concrete. Some are designed for use in new construction and are computed from data from construction details and environment, predicting time to first and subsequent repairs and life cycle costs for a new structure to compare durability options such as changing cement types, adding corrosion inhibitors, changing the concrete cover or changing bar types. Others are for projecting present data forward from existing structures to estimate time to damage.

Bazant developed a complete physical-mathematical model which describes the corrosion process in submerged concrete exposed to sea water (Bazant, 1979). A complete set of equations has been derived for the transport of oxygen and chloride ions through the concrete cover, the mass sources and sinks of oxygen and corrosion products, the cathodic and anodic electric potential, and the flow of ionic current through the concrete electrolyte. This model is largely based on theoretical assumptions and is completed by formulating the problem as an initial-boundary value problem which can be solved by using the finite element method. In order to arrive at numerical solutions, the spatial distribution and geometry of anodic and cathodic areas on the reinforcing bars have to be assumed. This model has been applied to several illustrative numerical examples. The results obtained show that for submerged concrete diffusivities for chloride ions and oxygen, not only at the anodic (rusting) area, but also, and mainly at, cathodic areas, usually are the controlling factors. The model takes into account all relevant chemical and physical processes involved in reinforcement corrosion, but several processes are not adequately addressed. Moreover, the polarization behaviour of the anodic and cathodic reactions has not been fully described. Despite these imperfections and the lack of experimental data, the model results in an improved understanding of the complex nature of the problem.

Similar models based on the same relationships, but expressed in more simplified forms, have been presented for example by Noeggerath (1990), Naish *et al.* (1990) and Raupach and Gulikers (1998). All these models have been applied to certain conditions showing that these analytical electrochemical approaches lead to suitable results for the specific corrosion problems to be analysed.

In November 1998, a workshop sponsored by the National Institute of Standards and Technology (NIST), the American Concrete Institute (ACI), and the American Society for Testing and Materials (ASTM), was held to discuss the different existing models for predicting service life of reinforced concrete exposed to chlorides and for estimating the life cycle cost of different corrosion protection strategies. The main outcome of that workshop was the decision to attempt to develop a 'standard' model under the authority of the existing ACI committee 365. A consortium was then established under the Strategic Development Council (SDC) of ACI. The consortium was established to develop an initial life cycle model based on the existing service life model developed at the University of Toronto (Boddy *et al.*, 1999). The life cycle model provides insight into the durability of a given construction practice, thereby allowing designers to ascertain the impact of additives to concrete and corrosion prevention strategies applied to new construction on the long-term durability of the concrete element. In October 2000, Life-365 version 1.0 software was released and version 1.1 followed in December 2001. The latest version at time of writing (2021) is version 2.2.3 which can be downloaded from the life-365.org website.

The Life-365 software predicts the initiation period assuming ionic diffusion to be the dominant mechanism. This software differs from other diffusion models in that it accounts for the variability of the diffusion coefficient with age and with temperature. It also attempts to model the impact of various additives. For additives such as silica fume and fly ash it reduces the diffusion coefficient to reflect the lower permeability and for corrosion inhibitors it raises the chloride threshold required to initiate corrosion. To include the impact of sealers and membranes it reduces the rate of accumulation of the surface chloride concentration. The rate of accumulation and the maximum accumulation of surface chloride in this program are based on the type of structure, geographic location and exposure. ACI Committee 365 has also published a state-of-the-art report on service life prediction ACI PRC-365.1-17: Report on Service Life Prediction.

The Sagüés's diffusion model, to predict the future performance of existing structures, has been validated on marine piles of several structures (Sagüés *et al.*, 1998; Sohanghpurwala and Diefenderfer, 2002). Initially, Sagüés used specialized software to calculate the diffusion coefficient from field cores and the solution to the error function to evaluate marine piles of two structures (Sagüés *et al.*, 1998). Later, Sohanghpurwala developed a methodology to perform the modelling in a standard spreadsheet program and validated the model on marine piles of another structure (Sohanghpurwalla and Diefenderfer, 2002). The Sagüés approach obtains probability distribution information on diffusion coefficient, clear concrete cover, and surface chloride ion concentration from the existing structure. It subdivides the concrete element into finite elements and determines the time to corrosion initiation for each finite element using the error solution to Fick's Second Law of Diffusion. To calculate the time to corrosion initiation (T_i) for each finite element, values of diffusion coefficient, clear concrete cover and surface chloride concentration are required. These are obtained from the probability distributions defined for each variable. The time for propagation (T_0) is assumed to be three to six years.

A similar software program has been developed in the United Kingdom. The diffusion coefficient is calculated from data taken from concrete blocks exposed to a marine tidal/splash zone environment, interpolated with laboratory data, and is used to estimate time to initiation (Bamforth, 1999). The time to failure is obtained from the Rodriguez formulae discussed earlier. Similarly, Broomfield *et al.* (2001) used the method of collecting chloride profiles from the structure under investigation to estimate the diffusion coefficient. The effective chloride diffusion coefficient was calculated using a parabolic approximation to the error function equation. From the individual sets of measurements, time to corrosion can be predicted. This is based on the actual diffusion and exposure characteristics of the measurement location. The Rodriguez equations are then used in a separate module of the program to predict time to cracking. The chloride diffusion model was validated against field data in one case where surveys were conducted approximately nine years apart. The earlier data set agreed with the later survey predictions of time to corrosion +2 years in 5 out of 6 cases where the initial prediction was greater than 10 years. In another 7 out of 9 cases they agreed corrosion was imminent in both cases but at the point of measurement there was no damage. There is also a model developed by Sohanghpurwala (2005), which was developed for chloride-induced corrosion of black steel and FBECR bridge superstructure elements. It also includes a generalized deterioration model that could apply to carbonation-induced corrosion.

References

Aligizaki, K.K. (2006). Concrete Cover Cracking as a Function of Rebat Corrosion: Theoretical and Experimental Studies. NACE 2006 Proceedings. 2006; Paper No. 06334.

ASTM C1202-18, 'Standard Test Method for Electrical Indication of Concrete's Ability to Resist Chloride Ion Penetration'. ASTM International, West Conshohocken, PA, 2018, 8 pp.

Bamforth, P. (2004). Enhancing Reinforced Concrete Durability. Technical Report 61. The Concrete Society, Camberley, UK.

Bamforth, P.B. (1999). 'The Derivation of Input Data from Modeling Chloride Ingress from Eight Year UK Coastal Exposure Trials'. *Magazine of Concrete Research*, 51(2): 87–96.

Bazant, Z.P. (1979). 'Physical Model for Steel Corrosion in Concrete Sea Structures – Application'. *Journal of the Structural Division, American Society of Civil Engineers*, 105(ST6): 1155–1166.

Boddy, A.B.E., Thomas, M.D.A. and Hooton, R.D. (1999). 'An Overview and Sensitivity Study of a Multi-Mechanistic Chloride Transport Model'. *Cement and Concrete Research*, 29: 827–837.

Broomfield, J.P. (2001). Trend 2000, BRE and Risk Review Ltd. Evaluation of Life Performance and Modeling, Building Research Establishment, Garston, UK.

Clear, K.C. (1989). Measuring Rate of Corrosion of Steel in Field Concrete Structures, Transportation Research Record, Transportation Research Board, Washington, DC.

Concrete Bridge Development Group Technical Guide 2 (2002). Guide to Testing and Monitoring the Durability of Concrete Structures. The Concrete Society, Camberley, UK.

Concrete Society/Institute of Corrosion (2004). Electrochemical Tests for Reinforcement Corrosion. Technical Report 60. The Concrete Society, Camberley, UK.

Feliu, S., Gonzalaz, J.A. and Andrade, C. (1994). Errors in the On-site Measurements of Rebar Corrosion Rates Arising from Signal Unconfinement. Concrete Bridges in Aggressive Environments. ACI SP-151 (ed. Richard E.W.), American Concrete Institute, Farmington Hills, MI.

Feliu, S., Gonzalez, J.A. and Andrade, C. (1995). Electrochemical Methods for On-site Determination of Corrosion Rates of Rebars, in Symposium on Techniques to Assess the Corrosion Activity of Steel in Reinforced Concrete Structures, December 1994, American Society for Testing and Materials.

Germann Instruments (1994). AlS RCT Profile Grinder Mark II Instruction and Maintenance Manual, Germann Instruments, Denmark.

Kropp, J. and Hilsdorf, H.J. (eds) (1995). Performance Criteria for Concrete Durability Rilem Report 12. E&FN Spon, London, UK.

Morinaga, S., Irino, K., Ohta, T. and Arai, H. (1994). Life Prediction of Existing Reinforced Concrete Structures Determined by Corrosion. Corrosion and Corrosion Protection of Steel in Concrete, Proceedings of International Conference held at the University of Sheffield, 24–28 July, 1994, R.N. Swamy (ed.), Sheffield Academic Press, 1994, pp 603–618.

Naish, C.C., Harker, A. and Carney, R.F.A. (1990). Concrete Inspection: Interpretation of Potential and Resistivity Measurements. Corrosion of Reinforcement in Concrete, International Symposium, Wishaw, Warwickshire, UK.

Noeggerath, J. Zur Makroelementkorrosion von Stahl in Beton: Potential- und Stromverteilung in Abhangigkeit vershiedener Einflubgroben. Eidgenossische Technische Hochschule, Diss, Zurich.

Parrott, L.J. (1987). A Review of Carbonation in Reinforced Concrete, C and CA Report for Building Research Establishment, Watford, UK.

Parrott, L.J. (1994a). 'Moisture Conditioning and Transport Properties of Concrete Test Specimens'. *Materials and Structures*, 27: 460–468.

Parrott, L.J. (1994b). Carbonation-Induced Corrosion, Paper presented at the Institute of Concrete Technology Meeting, Reading, 8 November, Geological Society, London.

Parrott, L. and Chen, Z.H. (1991). 'Some Factors Influencing Air Permeation Measurements in Cover Concrete'. *Materials and Structures*, 24: 403–408.

Poulsen, E. (1990). The Chloride Diffusion Characteristics of Concrete: Approximate Determination by Linear Regression Analysis, Nordic Concrete Research No. 1, Nordic Concrete Federation.

Purvis, R.L., Graber, D.R., Clear, K.C. and Markow, M.J. (1992). A Literature Review of Time-Deterioration Prediction Techniques, Strategic Highway Research Program, National Research Council, SHRP-C/UFR-92-613.

Purvis, R.L., Babaei, K., Clear, K.C. and Markow, M.J. (1994). Life-Cycle Cost Analysis for Protection and Rehabilitation of Concrete Bridges Relative to Reinforcement Corrosion, Strategic Highway Research Program, National Research Council, SHRP-S-377.

Raupach, M. and Gulikers, J. (1998). Determination of Corrosion Rates Based on Macrocell and Microcell Models – General Principles and Influencing Parameters. in EUROCORR 98 – Event No. 221, Utrecht, The Netherlands.

Ravindrarajah, R. and Ong, K. (1987). Corrosion of Steel in Concrete in Relation to Bar Diameter and Cover Thickness. ACI SP-100 Concrete Durability-Katherine and Bryant Mather International Conference, J.M. Scanlon (ed.), American Concrete Institute, Farmington Hills, MI, pp 1667–1677.

Rodriguez, J., Ortega, L.M. and Garcia, A.M. (1994). Assessment of Structural Elements with Corroded Reinforcement. Corrosion and Corrosion Protection of Steel in Concrete, Proceedings of International Conference held at the University of Sheffield, 24–28 July, 1994, R.N. Swamy (ed.), Sheffield Academic Press, Sheffield, UK, pp 171–185.

Sagüés, A.A., Scannell, W.T., Soh, F.W. and Sohanghpurwala, A.A. (1998). Assessment of Rehabilitation Alternatives for Bridge Substructure Components. Florida DOT.

Sohanghpurwala, A.A. (2005). Manual On Service Life of Corrosion Damaged Reinforced Concrete Bridge Superstructure Elements. NCHRP 18-06, A Final Manual.

Sohanghpurwala, A.A. and Diefenderfer, B. (2002). Project Development and Environmental Study SR64 from SR789 to East of Anna Maria Bridge. Florida DOT.

Tabatabai, H. and Lee, C.-W. (2006). Simulation of Bridge Deck Deterioration Caused By Corrosion. Proc. NACE Corrosion. Paper No. 06345.

Torres-Acosta, A.A. and Sagüés, A.A. (1998). Concrete Cover Cracking and Corrosion Expansion of Embedded Reinforcing Steel. Rehabilitation of Corrosion Damaged Infrastructure, Chapter IV: Modeling, Methods, Techniques and Technologies, P. Castro, O. Troconis and C. Andrade (eds), pp 215–229.

Torres-Acosta, A., Presuel-Moreno, F. and Andrade, C. (2019). 'Electrical Resistivity as Durability Index for Concrete Structures'. ACI Materials Journal, 116(6). 245–253.

Trend 2000. BRE; J Broomfield Consultants and Risk Review Ltd. Evaluation of Life Performance and Modelling. Corrosion of Steel in Concrete – Report D29. 2001 Dec; DTI DMI 5.1 (Report 2).

Tuutti, K. (1982). Corrosion of Steel in Concrete, CBI Swedish Cement and Concrete Institute, Stockholm.

Wang, X.M. and Zhao, H.Y. (1993). The Residual Service Life Prediction of R.C. Structures. Proceeding of the 6th International Conference on Durability of Building Materials and Components, S. Nagataki, T. Nireki and F. Tomosawa (eds), Omiya, Japan, October 25–29.

Weyers, R.E., Fitch, M.G., Larsen, E.P., Al-Qadi, I.L., Chamberlin, W.P. and Hoffman, P.E. (1994). Concrete Bridge Protection and Rehabilitation: Chemical and Physical Techniques – Service Life Estimates, Strategic Highway Research Program, SHRP-S-668, National Research Council, Washington, DC.

Chapter 10

Design for durability

10.1 Cover, concrete and design

Figure 10.1 shows what can happen in cases of inadequate design and maintenance of structure subject to chloride attack (Simon, 2004). Car parks are particularly vulnerable to corrosion. After several failures in the UK a guidance document has been produced on the inspection, maintenance and management of parking structures (Institute of Civil Engineers, 2002).

The first requirement for maximum durability against corrosion is low permeability concrete, with a high cement content, a minimal chloride content and good cover to the reinforcing steel. Good concrete and good cover should be designed in and enforced during construction regardless of other requirements and additional protection. Good cover and low water/cement

Figure 10.1 Collapse of the cantilevered section of a multi-storey car park in Syracuse, New York State due to deicing salt-induced corrosion (Simon, 2004). Courtesy Phil Simon.

DOI: 10.1201/9781003223016-10

ratios will increase the time taken for chlorides and carbonation to reach the reinforcing steel. All the major national and international design codes have tables specifying the cover and water/cement ratios needed for durability in different environments. Several years ago Bamforth (1994) stated that ordinary unblended Portland cements are inadequate for the most severe environments.

Additives and cement replacement materials such as pulverized fuel ash (fly ash), ground granulated blast furnace slag, silica fume (micro silica) and other materials can reduce the pore size and block pores enhancing durability further. However, proper curing may be vital to get the required performance with these blended cements.

The British Standard on design of concrete bridges, BS5400, Part 4, gave guidance on durability and environment, particularly concerning chloride environments. The nominal cover requirements for different corrosion environments are given for different concrete grades. However, the document *Design for Durability* (Department of Transport, 1994) increases these requirements by 10 mm. This specified increase may be due to the difference between nominal and actual cover. To ensure that 95% of the bars achieve an actual cover of 75 mm it may be necessary to specify a nominal cover of 85 mm. The document also notes that such high cover can lead to shrinkage cracking which may need to be controlled by adding fibres to the concrete.

The British Standards for construction are now superseded by the European Standard EN206 and its complementary British Standard BS8500. It is beyond the scope of this book to go into the details of the design codes and standards for reinforced concrete. EN206 is a complex document intertwined with several other documents. An overall guidance document for EN206-1 and BS8500 is available (BSI, 2015). One interesting aspect of the European standard is the exposure classes which are broken down into the classes given in Table 10.1.

This division of environments might have usefully been employed in BS EN 1504-9 (see Section 8.6), rather than or in addition to the addressing of anodic, cathodic and moisture control. In North America ACI 222.3R-11 (2003), 'Design and construction practices to mitigate corrosion of reinforcement in concrete structures', is a more succinct document that summarizes the design guidance in ACI 318 and the AASHTO bridge specifications. It also goes far wider than just mix design and cover, discussing corrosion inhibiting and other admixtures, coated and uncoated reinforcement, embedding other metals in concrete such as aluminium, lead, zinc, stainless steel, chrome, nickel and cadmium (coated steel). It also discusses steel placement, concrete consolidation, curing, coatings and barriers. It goes as far as discussing cathodic protection (sometimes referred to as cathodic prevention when referring to its application to new structures), and ECE, which would only be applied to new structures if chlorides had been cast in by mistake.

Table 10.1 Exposure classes from EN206-1

Class	Environment	Example
X0	Concrete with no embedded metal except at risk of freeze/thaw, abrasion or chemical attack or reinforced but very dry	Concrete in buildings with very low humidity, mass concrete
Carbonation risk		
XC1	Dry or permanently wet	Concrete in buildings with very low humidity, permanently submerged
XC2	Wet, rarely dry	Many foundations
XC3	Moderate humidity	External sheltered concrete
XC4	Cyclic wet and dry	External exposed surfaces
Corrosion induced by chlorides other than sea water		
XD1	Moderately humid	Exposed to airborne chlorides
XD2	Wet, rarely dry	Swimming pools, exposed to industrial water containing chlorides
XD3	Cyclic wet and dry	Bridges, car park slabs exposed to deicing salts
Corrosion induced by chlorides from sea water		
XS1	Exposed to airborne chlorides	Coastal buildings, etc.
XS2	Permanently submerged	Parts of marine structures
XS3	Tidal splash and spray zones	Parts of marine structures
Freeze thaw		
XF1	Moderate water saturation without deicing agents	Vertical surfaces exposed to rain and freezing
XF2	Moderate water saturation with deicing agents	Vertical surfaces exposed to rain, freezing and salt spray (e.g. bridge columns)
XF3	High water saturation without deicing agents	Horizontal surfaces exposed to rain and freezing
XF4	High water saturation with deicing agents	Horizontal surfaces exposed to rain deicing salts and freezing, for example bridge decks, marine splash zone

There is also an AMPP/NACE standard practice NACE SP0187 2017. It mentions stray-current-induced corrosion and corrosion monitoring systems which the other standards do not and has short paragraphs covering the same topics as ACI 222.3R-03.

When designing a structure, detailing and awareness of the climate and microclimate can avoid long-term problems. Installing waterproofing membranes on decks (in bridges and car parks), applying coatings or cladding to areas subjected to spray from vehicles or sea water will all delay the ingress of moisture and chlorides.

The designer should also make it as easy as possible to assess and repair the structure. One problem with post-tensioning is that there is no easy way to assess the condition of the tendons in the ducts. Even if we find corrosion how do we repair it? In the United Kingdom a committee was set up to answer the questions of design for durability, assessment and repair of post-tensioned, ducted concrete bridges. Their report has been published (Concrete Society, 1996). It deals with design, construction and quality systems for building new durable bonded post-tensioned structures. The problem of assessing the hundreds of existing structures that may be suffering from corroded tendons is now being undertaken by the Highways Agency.

The UK DoT *Design for Durability* Advice Note and Standard (1994) give clear guidance on minimizing the amount of reinforcement available for corrosion, maximizing access, improving drainage and minimizing runoff of salt laden water. Use of drips and proper guttering to control runoff is also discussed.

For structures with a very long lifetime requirement such as tunnels and other major civil engineering structures it is possible to carry out calculations to predict the likely time to initiation of corrosion for different concrete cover, mix designs and protective system (see Chapter 9). The chloride concentration at the surface can be predicted or estimated from structures in similar conditions. If special concrete mixes are being designed then the diffusion constant can be calculated according to the equations in Chapter 9 and the time for the corrosion threshold to reach the steel can be calculated. Calculating the consequences of reduced cover is also possible.

Other methods are available for enhancing durability by protecting the concrete surface, adding further protection to the concrete and protecting the steel. These are all discussed later. Another interesting issue is the installation of cathodic protection on new structures. This is also discussed at the end of the chapter. One problem with the standard discussed earlier is that there is no attempt to give the advantages and limitations of each protection system. The models in Chapter 9 will give life cycle cost information but not the pros and cons.

10.2   Fusion bonded epoxy coated rebars

The protection system of choice on highway bridges in North America for aggressive chloride conditions continues to be fusion bonded epoxy coated reinforcement (FBECR). It was first installed in a bridge deck in Philadelphia in 1973 (Manning, 1996). It was estimated by the FHWA that 100 million square feet (10 million m^2) of bridge deck on the US federal-aid highway system contains epoxy coated rebar up to 1889 (Virmani and Clemena, 1998).

There are several reasons for FBECR becoming the protective system of choice in the United States and Canada for reinforcing steel exposed to chloride attack. One is the reluctance to use waterproof membranes on bridge decks as they are difficult to install properly and to monitor both for correct installation and for performance after installation. The preference for a very low maintenance bridge deck led most state DOTs and the Federal Highway Administration (FHWA) to look for alternative protective systems on all bridges exposed to chlorides from the sea or from deicing salt.

In 1982 a panel consisting of David Manning (Ontario DOT), Ed Escalante (National Bureau of Standards) and the late David Whiting (Construction Technology Labs/Portland Cement Association) reviewed the performance of galvanized reinforcing steel. They recommended that it was not likely to give the long-term performance required (more than 40 years). This led to the US recommendation of FBECR over galvanized rebar (Manning *et al.*, 1982).

Although the corrosion inhibitor admixture calcium nitrite has been investigated and was considered acceptable by the FHWA, its performance in tests was deemed to be not as good as FBECR. The lack of competitive tender due to the patent being owned by a single manufacturer may also make states reluctant to specify it. Problems of flash setting and freeze-thaw damage have also been of concern when using nitrite inhibitors if the mix design is not adequately thought out. However, after their problems with FBECR (Manning, 1996) the Florida DoT is looking at micro silica concrete and calcium nitrite admixtures to achieve corrosion resistance along the coast line.

By a process of elimination as well as design, this has led to the preeminence of FBECR as the main protection of reinforcing steel from chloride attack in North American highway bridges, car parks and marine structures. In some Canadian provinces both waterproof membranes and FBECR are used. There is ongoing discussion in some state highway agencies about whether FBECR is the most cost-effective solution to bridge deck corrosion (Manning, 1996; Weyers *et al.*, 2002) but most states and provinces still use it for bridge decks exposed to deicing salts.

In Europe comparatively little FBECR is used, except in dowel bars for concrete pavements, where they have performed well for several years. They are also used for electrical isolation in tram and light railway system and for power transmission systems.

Compared with North America very few plants exist in Europe for manufacturing FBECR. The extensive use of waterproof membranes had minimized the problem of potholes on bridge decks. However, massive deck and joint repairs have been necessary over the last few years on many elevated sections of UK motorways due to chloride penetration through and round the ends of membranes. The first major European project to specify FBECR was the Great Belt Tunnel. This application is described in the next section.

10.2.1 How does epoxy coating work?

Epoxy coatings are applied to reinforcing steel in a factory process. The bar is grit blasted clean, it may then be pretreated and it is then heated in an induction furnace and passed into a coating unit that sprays a fine epoxy powder at the bar. The powder fuses onto the hot bar, is cured and quenched before passing out of the process.

The important property of the fusion bonded epoxy coating is that it is a dielectric, and charged species such as the chloride ion cannot pass through it. It also has excellent adhesion, properties to the steel and is not easily undercut by corrosion at defects. It must also be flexible to allow the straight bars to be bent during fabrication on a special mandrill to protect the coating from damage.

An alternative method of applying and using epoxy coated rebar has been used on the Great Belt Bridge/Tunnel lining segments in Denmark (Ecob *et al.*, 1995) and also the tunnel lining for the St Claire tunnel between the United States and Canada. In these projects the rebar cages are fabricated and the bars welded together. Instead of the coating being applied in a linear production line process the prepared cages are dipped in a fluidized bed of epoxy powder (Figure 10.2). Immediately after coating the cages can then be cast into the concrete with least risk of damage to the coating.

Defects are inevitable in any coating process. These come from several sources. There are pinholes in the coating after application because reinforcing steel is not smooth and has deformations on it. Small amounts of damage are done to the coating during handling and fabrication (bending) in the manufacturing plant. Some damage is inevitable during transporta-

Figure 10.2 A prewelded reinforcing cage emerging from the dipping tank on the Stoebelt bridge/tunnel prefabrication plant. Courtesy Mott MacDonald.

tion to site, and during the assembly of the reinforcement cages. Finally, the act of pouring wet concrete over the reinforcement and then vibrating it may cause further damage to the coating. The aim of all current standards on epoxy coated reinforcement is to keep the number and size of defects low and to repair them when possible. By keeping the number low, the size small and the separation high with a coating that minimizes undercutting corrosion there should be effective separation of anodes and cathodes (see Chapter 2). This increases the resistance in the corrosion cell and decreases the corrosion rate (Figure 10.3).

Epoxy coated rebars will corrode eventually if the chloride level builds up to a high enough level at the rebar surface and there is moisture and oxygen to fuel the corrosion reaction. However, they should take longer to start corroding than uncoated reinforcing steel and then should corrode more slowly than uncoated rebars.

10.2.2 Problems with epoxy coating

The problems found in the Florida Keys were first identified in the late 1980s. The Keys bridges were built in the late 1970s with FBECR. Routine inspection revealed localized corrosion in 1986 on V pier struts. Further corrosion was identified on a total of five bridges in subsequent inspections. The author visited the Keys in 1989 and saw extensive delamination in the tidal and splash zones with severe corrosion of the rebar exposed when delaminated concrete was removed with a hammer.

The following points should be noted when reviewing the Keys bridge corrosion problems.

1 Corrosion was localized to the splash and tidal zones.
2 There was no clear relationship between low cover and corrosion.
3 Although chloride levels were high, they were no higher than in non-corroding bridges in the Keys and elsewhere in Florida.
4 The other bridges in the Keys and 10 bridges elsewhere in Florida containing FBECR have been examined and are not yet showing corrosion problems. At least 20 bridges have been examined in detail.

It is now believed that poor handling practice with excessive exposure of coated bars to ultraviolet light, salt and mechanical damage had a role to play in the failures.

There is, however, a fundamental concern about the adhesion of the coating to the steel. FBECR is corrosion resistant because the epoxy coating is a dielectric film that stops charged ions (chloride) from passing through it. The other option of the ion is to go round it, for example, at defects or holidays in the coating. At these locations the ion should not be able to get under the coating to form a pit (see Chapter 2). A new, good quality FBECR

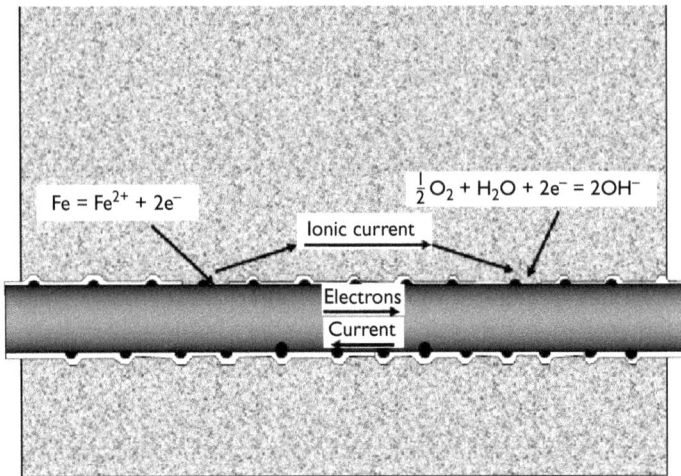

$$Fe = Fe^{2+} + 2e^-$$

$$\tfrac{1}{2}O_2 + H_2O + 2e^- = 2OH^-$$

Ionic current

Electrons

Current

Figure 10.3 Schematic of corrosion at anodic defect with ionic and electronic current flow to cathodic defect.

shows excellent adhesion of the coating to the steel and cannot be peeled off.

However, many bars in the Keys bridges, both those showing corrosion and those still in good condition, had poor adhesion and the coating could be peeled off the steel. There were also problems on new bars. An immediate programme of tightening application practice was undertaken. One major problem for adhesion is any deposits left on the surface after grit blasting and before coating. Salt is a particular problem as bars are often exposed to road and sea salt during transport to the coating plant and once in the grit used to clean the bars and provide a good key for the coating it can redeposit on the bar surface as the grit is recycled.

Even with good initial adhesion there was a loss of adhesion for concrete specimens exposed to regular or continuous wetting. Adhesion testing on the deformed surface of a rebar is difficult to perform quantitatively and there is some disagreement over whether the loss of adhesion when wet is permanent, fully or partially recovered if the bar dries out. Even if it is temporary, the bar is more susceptible to corrosion and under-film attack at defects while wet and once corrosion has started it will not recover adhesion.

Major research programmes have been undertaken in the United States and in other countries to find out whether the loss of adhesion is a fundamental problem in the durability of FBECR. If it allows extensive under-film corrosion such as that seen on the Florida Keys bridges then FBECR may not be a cost-effective solution to enhanced durability. Results are given in Clear *et al.* (1995), Virmani (1997) and Virmani and Clemena (1998).

All studies have shown that FBECR outperforms bare steel and galvanized steel in the laboratory and in the field. However, its detractors have suggested that the improved durability of field structures is associated with the reduced water : cement ratio of the concrete and increased cover that were implemented on all structures when using FBECR. For the foreseeable future it is likely that most North American highway departments will continue to specify FBECR and most European departments will not. Further tightening of specifications to avoid damage and to get the highest possible quality of coatings and coating adhesion have been implemented and may be improved further over the next few years. There is also research on how much damage occurs to the coating between leaving the coating plant and casting into concrete.

The position is changing and alternative corrosion protection methods such as the use of micro silica and calcium nitrite are being considered by Florida DOT and may spread to other major North American users of FBECR.

The use of prefabricated fusion dipped reinforcement cage should have long-term durability as the bending of the bars after coating has been seen to reduce adhesion and increase coating damage. It can be cost effective on large scale special projects. This was first used on the Great Belt Bridge/Tunnel in Denmark as described earlier. It has also been applied to the St Claire Railroad Tunnel between the United States and Canada. As the owners are looking for an almost indefinite life out of the structure, concrete mix designs were developed to minimize chloride diffusion.

10.2.3 Advantages and limitations of fusion bonded epoxy coated rebars

There are certain disadvantages to using epoxy coated rebars. Their insulating properties mean that they inhibit the use of electrical and electrochemical techniques. Taking reference electrode potential measurements on epoxy coated rebar can also be difficult. This is partly due to the electrical isolation of bars from each other as mentioned earlier, and partly because the steel is only exposed at small defects. The exposed steel at defects is not necessarily in an alkaline environment (either with or without chlorides) so interpretation of half cell potentials is difficult (see Section 4.7). It is therefore very difficult to interpret the measurements. For corrosion rate measurements with the linear polarization technique it is impossible to calculate the area of corrosion. For effective corrosion rate measurement the surface area must be known. If corrosion is only occurring at a few defects it may be impossible to determine the area of corrosion and therefore turn the corrosion current into an effective corrosion rate per unit surface area (see Section 4.11). Attempts have been made to take reference electrode measurements with some success, see for instance Hededahl and Manning (1989) and Ramniceanu et al. (2006).

There are problems with applying cathodic protection, chloride removal and realkalization to FBECR because the rebars are electrically isolated. Special care must be taken to connect each rebar together. This problem does not arise with the welded cages where each cage is continuous. The issues of repairing structures containing FBECR once they reach their limit state is described in Sohanghpurwala *et al.* (2002).

An important advantage of FBECR is that it is a comparatively low cost approach to corrosion protection. It is easily understood by site operatives and is now a routine technology in North America and some parts of the Middle East. If good quality rebars are cast into the concrete then there should be negligible maintenance requirements compared to waterproofing membranes or other surface coatings. There is no major impact for the bridge design other than an increase in lap lengths to take account of the reduced pull out strength of a bar with a deformable coating on it.

Special handling techniques must be used at the fabrication plant, in transit to the site and during handling and storage on site. It is essential that damage to the coating is minimized.

10.3 Galvanized rebar

As stated in Section 10.2, early US research generally showed galvanizing to be an inferior option to fusion bonded epoxy coated reinforcement when faced with chloride attack. The FHWA memorandum (Manning *et al.*, 1982) suggested a 15-year life for galvanized rebar in good quality concrete. Their conclusions are supported by Andrade *et al.* (1995) with respect to chloride attack. Galvanized bars may be more effective against carbonation. However, a recent re-appraisal of field structures suggests that it is more durable than previously thought in marine and other chloride environments. Yeomans (2004) gives numerous examples suggesting an increase in corrosion threshold from the range 0.2–1.0% for black steel to 1.0–2.0% for galvanized steel, reaching the 50-year life to first repair that the FHWA were looking for.

Galvanizing has one principal advantage over FBECR; it will accept damage in handling as the coating corrodes sacrificially and defects are not as important as for FBECR. Its main disadvantage is the risk of accelerated galvanic corrosion of the coating if galvanized reinforcement is linked to uncoated black steel.

Galvanized rebar is used successfully in structures where carbonation is a risk such as cladding panels. Galvanizing can easily be carried out in most countries although the quality and composition of the coating can affect its durability. It suffers from fewer problems when handled roughly because the coating is sacrificial and protects bare areas.

Accelerated depletion of the galvanizing can occur if galvanized rebars are mixed with ungalvanized bars. If they are being used in the same structure (for instance on a bridge deck but not the substructure or vice versa), then care should be taken to ensure complete electrical isolation of the galvanized and ungalvanized bars.

It may be very risky to apply electrochemical treatments to galvanized reinforcing steel. Very severe pitting can result. NCT, the patent holders on the realkalization and desalination techniques, do not recommend their use on structures containing galvanized rebar (Miller, 1995).

Galvanized steel reinforcement is probably one of the most widely used coated bar types after FBECR, although it is not widely used in Europe or North America. Its merits, limitations and use are comprehensively discussed in Yeomans (2004).[1]

10.4 Penetrating sealers

Several European DOTs specify the application of penetrating sealers to exposed concrete on bridges exposed to salt spray or run off. These are discussed in Section 6.3. They have very particular application requirements. In the United States sealers are also used on bare concrete bridge decks. Their effectiveness on trafficked surfaces has not been fully investigated.

The advantage of silane type penetrating sealers is that they are inexpensive compared with cathodic protection, epoxy coated rebars or other preventative or active treatment methods. If penetration is good then there is no maintenance and a long lifetime (in theory) as the silane forms a hydrophobic layer inside the concrete pores. There is no impact on the design or performance of the structure. The disadvantages are the problem of ensuring adequate application of a colourless liquid, and knowing that it has penetrated the structure.

Applying cathodic protection or chloride removal through penetrating sealers if necessary is usually possible.

10.5 Other corrosion resistant reinforcement

A comprehensive review of corrosion resistant reinforcement is published by AMPP/NACE NACE (2018)[2]. This gives a comparative review, under consistent headings, for a wide range of products, including stainless steels, stainless steel clad bars, low chromium steels, galvanised and other metallic coatings, and fibre reinforced polymer bars. It also covers stainless steel and epoxy coated prestressing strands as well as FBECR discussed above.

10.6 Waterproofing membranes

Waterproofing membranes are routinely applied to new bridge decks in the United Kingdom and much of Europe. Several DOTs in the Northern United States and Canada also apply them. Membranes come as two main types, a liquid that solidifies in place, and a sheet that is stuck to the concrete (see Section 5.3.3). The main problem with waterproof membranes is similar to that of epoxy coating rebars; they must be applied without defects. Defects can occur as blow holes, penetrations or mechanical damage to the liquid applied types, or cuts, tears, bad joins or perforations in the sheets. Either type of membrane can be damaged by the overlay of asphalt, either by its heat or mechanical damage from the aggregate particles.

Careful detailing is needed to avoid water and chlorides getting under the edges of the membrane. It is also essential to ensure that the deck and membrane drain properly, with adequate waterproofing around drains and with no ponding. The drainage system from the deck must not then dump salt laden water onto the substructure.

Many bridges in the United Kingdom (e.g. the Midland Links flyover and the Tees viaduct) have required extensive concrete repair to the bridge decks where membranes have failed to stop water and chloride ingress. The repair problems are worse at joints, where the membrane ends. The sealing of joints and making membranes continuous across them is a popular solution to this problem. Modern bridge design requires the minimizing of the number of joints in a bridge deck. There must be adequate detailing and proper attention to ensuring that the water does not accumulate and that membranes run into drains and over concrete edges where salt might accumulate. A synthesis report ton bridge deck joint performance in the United States has recently been undertaken by Purvis (2005).

Problems are also found at curbs, around drains and other places where the salt water can get under the membrane.

Waterproofing membranes have a life of about 15 years and then must be replaced. At that time it is good practice to survey for corrosion and chloride penetration after the membrane has been exposed. One problem with membranes is determining whether they even exist under the asphalt wearing course, or whether they are still working. It is sometimes possible to flood the deck with water and measure for electrical continuity between the surface and the reinforcing steel. This is difficult and will not find the defect. SHRP carried out research on using pulse velocity to define a condition rating for an asphalt overlaid membrane. This showed promise but needs further research before it can be developed into an accurate test.

The use of waterproofing membranes requires careful detailing of a structure at the design stage with good drainage and effective prevention of salt laden water getting under the membrane. This must be carried through to the construction site. A good finish must be established for the membrane to be applied successfully without debonding. Application of membranes requires a specialist to ensure there are no defects and that the membrane is adequately bonded to the concrete and sealed at the edges. The asphalt wearing course must be carefully placed to prevent damage to the membrane. The complete design of the structure must take into account the extra dead load and clearance required for the asphalt.

Some waterproofing membranes on car parks do not require asphalt concrete overlays. The extra wear on these membranes, especially from scuffing of tyres as cars manoeuvre in and out of parking spaces, means that they have shorter lifetimes and need more regular monitoring and maintenance if they are to keep the deicing salts out of the concrete.

10.6.1 Advantages and limitations of waterproofing membranes

If membranes fail, detecting the failure is difficult as they are overlaid with an asphalt wearing course. Leakage can occur through thinner slabs such as those on car parking structures, and traffic will degrade the asphalt overlay which will require replacement at regular intervals. There can be particular problems in car parks due to the manoeuvring of cars into parking spaces and round tight corners displacing the asphalt and damaging the membrane. Car park membranes often combine the waterproofing layer and the wearing course in one. Where a car park is built over offices or shopping malls leaking membranes can cause considerable problems.

Membranes have given good service on European bridges and in the New England states and other North American agencies where they are routinely used. They have reduced or eliminated the problems of potholing on bridge decks but at the price of a higher regular maintenance requirement compared with epoxy coated rebars in bridge decks.

Until recently membranes were not considered compatible with cathodic protection as the anode would have to be under the membrane to pass current, and the anode generates gases that must escape. Membranes and cathodic protection have been used in car parks on thin slabs with some success. Some of the newer probe anodes have venting tubes and so they can be used under membranes. Galvanic anodes can be used as they do not generate gasses and electrochemical chloride removal (or realkalization) could be done before the replacement of the membrane.

10.7 Stainless steel reinforcement

Stainless steel rebar has been applied in special circumstances but it is a very expensive option. A clad stainless steel with a mild steel core and stainless outer 1–2 mm is presently being commercialized, which is still relatively expensive, but about half the price of pure stainless steel reinforcing bar. Stainless steels can be susceptible to pitting attack, so the correct grade of stainless steel must be used. There appear to have been problems of quality control and assurance of the manufacturing process (Pianca, 2001).

The problem of mixing metals applies to stainless steel with mild steel as with galvanized and ungalvanized steel in the section earlier. The galvanic coupling of mild steel and stainless steel will accelerate corrosion of the mild steel. This problem has been observed on balconies and façades (Miller, 1995). However, guidance documents on stainless steel by the Concrete Society (1998) and the European Federation of Corrosion by Nürnberger (1996) indicate that as stainless steels are poor cathodes the galvanic effect is minimal and there is little acceleration of the corrosion rate. It is obviously good design practice to ensure that the mild steel reinforcement has a very low risk of exposure to chlorides if being mixed with stainless steel bars.

With the development and use of low alloy stainless steels it becomes more important to define the threshold for corrosion. This is discussed by Kessler (2022), who reviewed available test techniques. However, the best test appears to still be under development.

10.8   Corrosion inhibitors

Calcium nitrite is the principal corrosion inhibitor available to stop corrosion that is compatible with concrete in the casting process. As stated in Section 10.2, it is accepted by FHWA as an alternative to FBECR for protection against chloride-induced corrosion. FHWA research shows that if sufficient nitrite is added to the concrete mix to ensure a chloride to nitrite ratio of less than 1.0 at rebar depth, then the nitrite will prevent corrosion. Obviously this is feasible in marine conditions where the chloride level is known (and assuming no concentration effects) but may be more difficult in other situations.

Compared to FBECR an inhibiting admixture has the advantage that there is electrical continuity of the reinforcing cage so electrochemical methods can be used to measure for corrosion measurements (reference electrode potential measurements and corrosion rate) and to treat for corrosion (cathodic protection and chloride removal). Its major drawback is that with a good quality, dense concrete (Florida DoT now specifies micro silica in the concrete) and good cover, the inhibitor should not be needed for at least 20 years. Will it still be there and will it still be effective 20 years after being put in the concrete mix?

The advantage of calcium nitrite is that it can be added to the mix and has no serious effects on the design, construction and performance of the structure other than its effect as a set accelerator. Mix design may require adjusting to include a retarder. Its disadvantage is that there must be enough to stop corrosion and it is consumed with its exposure to chlorides. It is therefore important to calculate the chloride exposure for the life of the structure and add sufficient inhibitor. It does not inhibit the application of cathodic protection or chloride extraction in later life of the structure if necessary.

Other proprietary corrosion inhibiting admixtures, mainly amine based, are also available although there is much discussion about longevity and effectiveness. Some of these issues are covered in Elsener (2001). As an additive to the concrete mix, inhibitors can be used along side other protective measures such as coated reinforcement.

10.9 Installing cathodic protection in new structures

There are two reasons for installing cathodic protection from new. One is to energize it initially as a 'cathodic prevention system'. The other reason is to have it there for use at a future date.

The first approach has been pioneered on Italian Autostrada bridges in the mountainous areas of Northern Italy (Pedeferri, 1992). Previous construction experience had shown that it was very difficult to prevent chloride ingress. Therefore cathodic protection anodes were either built into the deck or applied as part of the construction process. The bridges were post-tensioned segmental construction with tendons in grouted ducts.

Cathodic protection was applied at a very low level to avoid any risk of hydrogen evolution. As chlorides had not entered the structure the aim was to apply a small level of current to polarize the steel and effectively repel the chlorides. These systems were installed in the late 1980s and early 1990s. If properly maintained they should give excellent life as the anodes will be used at a very low rating. Another case of installation from new was a salt silo or chamber with a reinforced concrete floor. This was known to be at high risk from chloride penetration. Further examples are a palace in the United Arab Emirates (Funahashi, 1995) and a petrochemical plant in Saudi Arabia (Chaudhary, 2004).

Although not much used in Europe or north America, it is more widely used in the Gulf states due to the aggressive condition and prevalence of chloride. One of the largest recent projects is the Louvre, Abu Dhabi, where conventional buried submerged CP technology was used to protect the 5000 steel piles supporting the building complex and mixed metal oxide coated titanium ribbon system were applied to approximately 80,000 m^2 of basement slabs, external walls and other areas exposed to saline conditions. The design life is 100 years. ICCP is now routinely applied during construction to industrial plant in Saudi Arabia where chloride exposure is high.

The most difficult task with installing anodes during construction is ensuring isolation between the anode ribbons and steel. Once the concrete is cast, repair is probably impossible and the constraints of the construction schedule, often with overnight concrete pours in the Gulf states, make it difficult to ensure isolation is adequately monitored and controlled. The use of special clips to keep anodes away from bars is strongly recommended. Alternatively, for ground slabs, putting the anode in the blinding layer prior to installing the cage can minimize the risk of short circuits.

The installation of anodes and rebar connections during construction for later use has been discussed on a number of structures. While positive efforts to ensure rebar continuity at the construction stage are to be welcomed, it is generally considered that anode design and types will change over the next decade so improved systems may be available by the time that protection is required. If corrosion is local then the application of small, local anodes at the time of corrosion may be more cost-effective than designing in anodes to protect the whole of the structure. Also if there are inaccessible locations that are at risk on a high value structure where access or outage costs are high then installing cathodic prevention may be a worthwhile precaution. The author has been involved in such an installation on a desalination plant and other seawater intakes where shut down costs would be prohibitive.

Experience has also shown that connection leads get damaged and destroyed or may not be in the correct location for future use. If an impressed current cathodic protection system is to be installed from new then it should be commissioned and energized before hand over to ensure that it is working and to ensure that the system is fully operational and documented.

10.10 Durable buildings

Much of this chapter has discussed bridges, with some discussion of buildings and other civil engineering structures. This is partly because most of the emphasis in the field of durability has been on problems with bridges caused by chloride attack. Many of the discussions are relevant to all reinforced concrete structures. However, buildings are constructed to withstand less severe environments and conditions than bridges. There are also lower life expectations on buildings which may become obsolete before serious deterioration occurs.

The trend towards covering up concrete with coatings and cladding increases the durability of the concrete by retarding chloride and carbon dioxide ingress. Buildings rarely use FBECR, micro silica, sealers or inhibitors, but they rarely suffer from severe chloride exposure.

If sufficient attention is paid to ensuring the specified cover and water cement ratio, most buildings in relatively benign environments will perform well throughout their normal lifetimes. In aggressive environments such as many countries in the Middle East or where exposed to sea salt spray, designers should consider the durability codes for bridges rather than those for buildings. The CIRIA guide to concrete construction in the Gulf Region, CIRIA (1984), is also a good guide.

Adequate maintenance and maintainability are also important. If leaks develop on the flat roof of a building exposed to the sea then chlorides can concentrate and cause damage. If drainage is well designed and maintain then problems will be minimized.

Table 10.2 Methods of improving durability (Adapted from Broomfield, 2004)

Method	How it works	Advantages	Limitations	Where used
Increase cover over reinforcement	Increases time for chlorides or carbonation to reach steel	Low cost, included in codes	Requires very good quality assurance, increased risk of cracking giving Cl^- and CO_2 more rapid ingress	Everywhere
Decrease water/cementitious ratio	Reduces concrete porosity to increase time for corrodents to reach steel	Low cost, in codes	Makes concrete less workable and more prone to poor consolidation, higher cement content increases heat of hydration. Plasticizers add to cost	Everywhere
Ground granulated blast furnace slag, microsilica, fly ash	Reduces concrete porosity to increase time for corrodents to reach steel	Modest cost, increasingly in codes to achieve durable concrete in marine exposure lower cement content decreases heat of hydration	Makes concrete less workable and more prone to poor consolidation, costs and availability can be problematical Availability will decrease with decreased use of coal and coke in power stations and smelters	Marine Structures and other severe exposure
Corrosion inhibitors	Protective chemical layer on steel	Modest cost, can be used with other techniques	Needs to work after 20 years or more, may be washed out in marine environment. Dosage must reflect exposure	Marine structures, car parks and other severe exposure
Coatings and penetrating sealers	Increases time for chlorides or carbonation to reach steel	Low cost, easy maintenance	Changes appearance requires maintenance. Application of sealers weather dependent	Bridges, car parks
Membranes and barriers	Increases time for chlorides or carbonation to reach steel	Ideal for bridge, car park decks etc.	Requires maintenance, can hide problems. Changes profile and dead load	Bridge and car park decks
Fibre reinforced polymer reinforcement	Corrosion resistant	Theoretical very long life. Easy handling, may reduce concrete specification	Needs new design codes, unknown deterioration rates and mechanisms over 50y life or more	Decks and flooring
Stainless steel reinforcement (solid or clad)	Corrosion resistant reinforcement	If correct grade is used, indefinite life. Scrap value of solid bar	High first cost. Clad bar production QA/QC still in development	High value long life structures or elements
Cathodic prevention	Makes steel cathodic and therefore corrosion resistant	Theoretically up to 100y life of anodes. Codes avaiable	High first cost, high maintenance, technically complex	High value high chloride exposed structures
Epoxy coated reinforcement	Coating prevents chloride access to steel surface	Easy low cost use in North America	Durability problems in some highway structures in Florida and Ontario. Handling damage during transportation and placement	Bridge decks, etc. in North America and Arabian Gulf area
Galvanized reinforcement	Higher chloride tolerance and resistance to carbonation delays corrosion, sacrificial protection	Widely available, design, handle and use as black steel	Advantages reduce in poor concrete and high chloride. Risk of accelerating corrosion if coupled to black steel	Everywhere

10.11   Conclusions

There are many methods of ensuring that reinforced concrete structures do not suffer from corrosion problems. The most important is ensuring that the concrete quality is high, with a low permeability, and the cover is good. Additional protection can be achieved by keeping chlorides away by design, such as removing joints in decks where they can allow chlorides to leak on to substructures or applying coatings or cladding to façades of buildings exposed to sea salt spray. Further options are offered in the way of surface barriers, coatings, membranes and sealers. These can be used for a variety of applications to protect specific elements or specific areas at risk. Additional protection can be found in the form of blended cements with pozzolanic additives that can retard ingress, such as micro silica, PFA (fly ash), etc. The addition of corrosion inhibitors to the mix can be included in this category, although it bridges the distinction between a barrier to corrodents and a technique offering corrosion protection to the steel.

The earlier techniques were concerned with stopping chlorides and carbonation reaching the surface of the steel. The next category is making the steel corrosion resistant. These include epoxy coating, galvanizing and stainless steel bars. The problems with the former are performance. The problem with the latter is cost. Coatings such as epoxy or zinc impede our ability to assess and repair damage when it does occur. Finally there is cathodic protection that can be applied from new and is being used on several new structures to stop chlorides reaching the bar and to stop the bar from corroding if the corrodents reach it. A paper by Atkins *et al.* (2021) reports that the application of ICCP to a new reinforced concrete structure in an aggressive environment had a lower embodied energy than other options for extending the design life. Table 10.2 summarizes the techniques and their pros and cons.

Notes

1 (Ed), Y. (2004). 'Galvanized steel reinforcement in concrete'. Galvanized Steel Reinforcement in Concrete Hardcover ISBN: 9780080445113 Imprint: Elsevier Science Published Date: 26th November 2004
2 NACE (2018) TR21429 State of the Art Report on Corrosion Resistant Reinforcement, AMPP, Houston
3 Atkins, C. P., et al. (2006). Sustainability and Repair. Concrete communications conference. TY – JOUR.

References

Andrade, C., Hoist, J.D., Nürnberger, N., Whiteley, J.J. and Woodman, N. (1995). *Protection Systems for Reinforcement*, Prepared by Task Group VII/8 of Permanent Commission, VII CEB.

Atkins, C. & Lambert, P. (2021). Sustainability and Corrosion. Proceedings of the Institution of Civil Engineers – Engineering Sustainability. 1–13. 10.1680/jensu.21.00011.

Bamforth, P. (1994). 'Admitting that Chlorides Are Admitted'. *Concrete*, 28(6): 18–21.

Broomfield, J.P. (2004). Galvanized Steel Reinforcement: A Consultant's Perspective. Chapter 9 in Yeomans (ed.). Galvanized Steel Reinforcement in Concrete, Elsevier, Oford UK, pp 276–279.

BSI (2015). BS 8500-1:2015+A2:2019 Concrete. Complementary British Standard to BS EN 206. Part 1: Method of Specifying and Guidance for the Specifier. British Standards Institute, London.

Chaudhary, Z., Bairamov, A.K., Rodel, F. and Al-Mutlaq, F. (2004). 'Cathodic Protection of Reinforced Concrete Seawater Structures in Petrochemical Plants'. *Materials Performance*, 43(12): 26–29.

Clear, K.C., Hartt, W.H., McIntyre, J. and Seung, K.L. (1995). Performance of Epoxy Coated Reinforcing Steel in Highway Bridges, National Cooperative (Highway Program, Report 370, National Academy Press, Washington, DC.

Concrete Society (1996). Durable Bonded Post-Tensioned Concrete Bridges. Technical Report No. 47. The Concrete Society, Camberley, UK.

Concrete Society (1998). Guidance on the Use of Stainless Steel Reinforcement. Technical Report No. 51. The Concrete Society, Camberley, UK.

Department of Transport (1994). Design for Durability, Departmental Advice Note. BA 57/94.

Ecob, C.R., King, E.S., Rostam, S. and Vincentsen, L.J. (1995). Epoxy Coated Reinforcement Cages In Precast Concrete Segmental Tunnel Linings – Durability. Corrosion of Reinforcement. In Concrete, C.L. Page, K.W.J. Treadaway, P.B. Bamforth (eds), published for the Society of Chemical Industry by Elsevier Applied Science, London and New York, pp 550–558.

Elsener, B. (2001). Corrosion Inhibitors for Steel in Concrete. European Federation of Corrosion Publications, Institute of Materials, London, p 35.

Funahashi, M. (1995). 'Cathodic Protection Systems for New RC Structures'. *Concrete International*, 17(7): 28–31.

Hededahl, P. and Manning, D.G. (1989). Field Investigation of Epoxy-Coated Reinforcing Steel. Ontario MTO Report. MAT-89-02.

Institute of Civil Engineers (2002). Recommendations for the Inspection, Maintenance and Management of Car Park Structures, ICE, London, UK, ISSN: 0 7277 3183 1.

Kessler, S. The Determination of Corrosion Threshold of Stainless Steel in Concrete - a Review, Paper 16179 AMPP/NACE Corrosion 2022, Houston, TX.

Kessler, S. (2021). The Determination of Chloride Thresholds of Stainless Steel in Concrete – A Review, Paper 16179, AMPP Corrosion 2021

Manning, D.G. (1996). 'Corrosion Performane of Epoxy-Coated Reinforcing Steel: North America Experience'. *Construction and Materials*, 10(5): 349–365.

Manning, D.G., Escalante, E. and Whiting, D. (1982). Panel Report – Galvanized Rebar as a Long-Term Protective System. FHWA- DTFH61-82-300-30041-2/3.

Miller, J. (1995). Personal Communication.

Nürnberger, U. (1996). Stainless Steel in Concrete – State of the Art Report. Institute of Materials for EFC, London, ISBN: ISSN 1354-5116.

Pedeferri, P. (1992). Cathodic Protection of New Concrete Constructions, in International Conference on Structural Improvements through Corrosion Protection of Reinforced Concrete, Documentation E7190, IBC Technical Services, London, VK.

Pianca, F. (2001). Assessment of Stainless Steel Clad Reinforcement for Ministry Use. TRB Presentation. Protection Systems for Reinforcement, Prepared by Task Group VII/8 of Permanent Commission VII CEB.

Purvis, R. (2005). Bridge Deck Joint Performance. NCHRP Synthesis 319.

Ramniceanu, A., Weyers, R.E., Anderson-Cook, C. and Brown, M.C. (2006). 'Measuring the Field Corrosion Activity of Bridge Decks Built with Bare and Epoxy Coated Steel'. *Journal of ASTM International*, 3(8).

Richard, E.W., Prof. Brown, M.C., Kirkpatrick, T.J., Weyers, R.M., Mokarem, D.W. and Sprinkel, M.M. (2002). Field Assessment of the Linear Cracking of Concrete Bridge Decks and Chloride Penetration. Transportation Research Board 81st Annual Meeting CD ROM. 81st Annual meeting proceedings. Paper No. 3654.

Simon, P. (2004). Improved Current Distribution due to a Unique Anode Mesh Placement in a Steel Reinforced Concrete Parking Garage Slab CP System. NACE Corrosion 2004. Paper No. 04345.

Sohanghpurwala, A.A., Scannel, W.T. and Hartt, W.H. (2002). Repair and Rehabilitation of Bridge Components Containing Epoxy-coated Reinforcement. NCHRP Web Document 50 (Project 10-37C). Contractor's Final Report.

Virmani, Y.P. (1997). Corrosion Protection Systems for Construction and Rehabilitation of Salt-contaminated Reinforced Concrete Bridge Members. Proceedings of the International Conference on Repair of Concrete Structures – From Theory to Practice in a Marine Environment, Svolvær, Norway, pp 107–122.

Virmani, Y.P. and Clemena, G.G. (1998). Corrosion Protection – Concrete Bridges. FHWA Report. FHWA-RD-98–088.

Yeomans (ed.) (2004). Galvanized Steel Reinforcement in Concrete, Elsevier, Oxford, UK.

Chapter 11

Sustainability and future developments

About 15 years ago the author wrote the corresponding chapter in the second edition of this book. It is interesting to see what was predicted in that edition, what has actually transpired and what we might expect to transpire over the next decade or two compared with the last.

As stated at the beginning of the book, concrete is man's most ubiquitous invention, with more concrete made than any other man-made material. It is not surprising therefore to read of new alternatives to concrete made from oil residue, a rapid carbonating concrete to 'mop up' carbon dioxide and other variations to challenge the supremacy of Portland cement concrete and its related hydraulic cement.

The construction industry is a conservative one compared with others such as microprocessors and computers. However, with a design life of a few years, hardware and software is mass produced, upgraded and rapidly superseded. The construction industry does not have that luxury. Our products are expected to last many decades and in many cases for centuries. Until recently, there was little reference to the ultimate design life in the specification, design and construction of many structures. However, that is changing, a requirement made more important with concern about climate change and the embodied energy of existing structures and new build, more thought is being given to ensuring better durability. Concrete applications range from simple paths and blocks to the tallest buildings in the world and environments ranging from the most benign indoor exposure to offshore platforms in the North Sea. Therefore, it is not surprising to read of new alternatives to concrete made from oil residue, a rapid carbonating concrete to 'mop up' carbon dioxide and other variations to challenge the supremacy of Portland cement concrete and its related hydraulic cement. Some may have niche applications but it will take a lot of research and development work to find a new cement and concrete that will challenge the durability, cost effectiveness and compatibility of Portland cement-based concrete.

In the first edition we suggested that admixtures to enhance the durability of concrete would continue to be developed. The major one that is having an impact on new construction seems to be superplasticizers which are

DOI: 10.1201/9781003223016-11

eliminating the problems of consolidation around the reinforcement and producing dense, high quality concrete even in the most complex formwork and reinforcement design both in new construction and in repairs. The latest innovation is the move toward 3D printing of concrete. Recent articles in the technical press discuss the use of carbon nanotubes for reinforcement, challenging current standards and codes of practice, given the difficulty of using conventional steel reinforcing bars.

The issue of the environment and environmental impact has been given additional impetus by global warming with articles in the technical press about 'zero net carbon' structures in the past 10 years for many parts of the construction industry. Highway agencies continue to move forward using recycled materials and waste products in concrete. Concrete repair itself could be viewed as an environmentally friendly approach as the alternative would be new construction with the transportation of large amounts of material both for demolition and reconstruction as well as the impact of manufacturing the new cement, and steel and quarrying and crushing the aggregates. Waste disposal in landfill sites is increasingly discouraged, at least in the United Kingdom, so prices of disposal rise, encouraging life extension and recycling.

The present drive to control global warming means that there is increasing emphasis on controlling energy use in general and CO_2 output in particular in new build and repair. In a recent paper Atkins and Lambert (2021) discussed the embodied energy and embodied carbon (CO_2) equivalent of conducting repairs and preventative maintenance. They looked at coatings on steel, coatings on concrete at risk of carbonation-induced corrosion and concrete repairs and impressed current cathodic protection applied to reinforced concrete structures at risk of chloride-induced reinforcement corrosion. They found that impressed current cathodic protection has a lower embodied CO_2 than repeated patch repairs and cathodic protection from new can have a lower CO_2 equivalent than repairs in a chloride exposure environment.

We can expect their methodology to be refined for wider use and similar calculations to be required for specific new build and repair projects. Values for the embodied CO_2 equivalent will become more readily available and more accurate as the requirement for such figures increase. This should lead to an increase in structure preservation where the whole life cost and the whole life energy and embodied CO_2 impacts choices about repair and rehabilitation versus demolition and new build. This of course will depend upon the CO_2 equivalent emissions of the rehabilitated structure compared to a replacement, especially for buildings.

The high price of metals on the international markets has held back wider use of stainless steel at the time when its use was starting to be more widely considered. The low chrome corrosion resistant steels discussed in NACE (2018) are being used in the USA and Canada but have had limited impact elsewhere. The move to develop new organic coatings for reinforcement seems to have diminished as no clear improvement to fusion bonded epoxy coating has yet emerged.

The experiments with alternatives to steel reinforcement continue, such as glass fibre composites, but these require major changes to codes so are unlikely to make a major impact quickly. Carbon fibre and other composite wraps are increasingly being used in strengthening repairs. The author is still concerned that they are also being mooted for corrosion control, despite the potential hazard of failure without warning seen for conventional concrete jacketing as shown in Figure 6.10.

The electrochemical techniques continue to develop as predicted previously. Galvanic cathodic protection is now a commercial reality and new anode systems are in development. There are suggestions that impressed current and galvanic systems may be combined with a short burst of initial impressed current and then leaving the galvanic anodes to maintain polarization (Davison et al., 2006). The range of impressed current anodes for cathodic protection continues to grow along with interest in wider application of the technology such as steel framed buildings. Surprisingly realkalization and ECE have not expanded in the same way as cathodic protection. Electro-osmosis is in its infancy as a corrosion control technology. It will either continue to develop or find its niche over the next decade.

In North America the market for impressed current cathodic protection of bridge decks diminished (Russell, 2004) although galvanic systems are now being preferred due to lower maintenance. High-performance concrete overlays, full and partial deck replacement are preferred due to the cost of maintenance.

Also as predicted there has been little change in deicer technology. Some US states are using calcium chloride instead of sodium chloride and there is concern about the environmental impact of runoff into water courses in sensitive areas so there is more use of grit alone or mixed with salt.

Corrosion monitoring systems (Chapter 5) are gaining acceptance and use around the world with the continued reduction cost of electronics and the increase in computing power it is easier to collect and analyse data. There is considerable research on wireless systems in the United States.

The issue of the durability of prestressed structures has not gone away. Although the United Kingdom published its guidance document on durable prestressing (Section 10.1), there are now problems in Florida with bleed water from the grout accumulating at the anchorages and leading to corrosion (Sagüés et al., 2005) exacerbated by chloride contamination of the grout, see FHWA 2011. This is likely to lead to further research and developments of grouts of post-tensioning and of anchorage design.

There has been a huge growth in models for corrosion of steel in concrete with the ability to do whole life costing on new structures incorporating different durability options and on repair options for existing structures

(Chapter 9). This has the big advantage of codifying our approach to durability. Although all such models should be considered comparative rather than absolute in their output the client can now be presented with engineering judgement rather than prejudice and preference for known and tried techniques, some of which may not be very good. It will be interesting to see what models are developed for calculating embodied CO_2 for constructions, rehabilitation, corrosion and whole of life.

Perhaps the biggest growth area over the past years is the growth of codes and standards. The European Union's attempts to generate a common market with common standards has driven some of that, although in the USA the new AMPP, combining NACE and SSPC lists 35 documents on corrosion and concrete, either under development or being maintained. There is a need for codification of the techniques we use although they are no substitute of intelligent design and engineering of a structure. As stated previously, manufacturers need performance standards and specifications so that they have fixed goal posts for the properties and performance of the products they develop.

We will always need to build structures in corrosive environments. We will therefore need new techniques to deal with the problems that ensue. We will also continue to refine the techniques that we have. We will continue to need engineers and corrosion specialists with knowledge and experience of dealing with the problems as they are the most important resource in the fight against corrosion. Good undergraduate and post graduate courses on corrosion are needed as well as research projects to train young engineers in the theory and practice of corrosion of steel in concrete. Perhaps the biggest challenge is to guarantee that those responsible for the design of new structures ensure that there is sufficient understanding of durability in the design team so that many of the problems we see today are designed out of the structures of tomorrow.

References

Atkins C. and Lambert P. (2021). 'Sustainability and Corrosion'. *Proc Instn. Civil Engs*, Engineering Sustainability, 1–13.

Davison, N., Glass, G. and Roberts, A. (2006). Inhibiting Chloride Induced Reinforcement Corrosion. Proceeding NACE Corrosion. Paper No. 06353.

FHWA (2011). Memorandum Elevated Grout Levels in SIKA 200 PT Cementitious Grout, Federal Highway Administration, US Department of Transportation, McLean, VA.

Russell, H.G. (2004). Concrete Bridge Deck Performance: A Synthesis of Highway Practice. NCHRP Synthesis. 333.

Sagüés, A.A., Wang, H. and Powers, R.G. (2005). Corrosion of the Strand – Anchorage System in Post-Tensioned Grouted Assemblies. NACE Corrosion 2005. Paper No. 05266.

Appendix

Sources of information on corrosion of steel in concrete

Bodies involved in the corrosion of steel in concrete:

American Association of State Highway & Transportation Officials
Web: www.transportation.org
Publishes AASHTO/ARTBA/AGC Guide Specifications for reinforced concrete construction and repair of highway structures

American Concrete Institute (ACI)
Web: www.concrete.org
Publishes guidance documents on reinforced durability, modelling and repair. Annual Conference training and certification in concrete construction. Journals ACI Journal and Concrete International.

American Society for Testing and Materials (ASTM)
Web: www.astm.org
Publishes test standards for concrete and corrosion, seminars and conferences.

Australasian Corrosion Association
Web: http://www.corrosion.com.au
Annual conference on corrosion. Training and certification courses in corrosion

British Standards Institute
Web: shop.bsigroup.com
Publishes British and European standards on corrosion, concrete and concrete repair.

Corrosion Prevention Association (CPA)
Web: www.corrosionprevention.org.uk/
Collaborate with Concrete Society and ICorr on Concrete Society Reports, Technical reports on repair techniques, particularly electrochemical. Runs ISO certification courses for CP of steel in concrete for ICorr.

Concrete Society
Web: www.concrete.org.uk
Publishes 'Concrete' and 'Concrete Engineering International'. Annual Conference, Concrete Society Technical Reports

Concrete Repair Association (CRA)
Web: www.cra.org.uk
Guidance document. Website information on EN1504

European Federation of Corrosion (EFC)
Web: www.efcweb.org
Annual Conference, publishes reports on corrosion of steel in concrete via IOM^3

Institute of Corrosion
Web: www.icorr.org
Publishes report on corrosion of stel in concrete, ofen in collaboration with the Concrete Society and recently with CPA.
Training and certification courses in corrosion

The Institute of Materials, Minerals and Mining (IOM^3)
Web: www.iom3.org
Published European Federation of Corrosion documents

International Concrete Repair Institute
Web: www.icri.org
Annual conference, reports and guidance documents for concrete repair

Association for Materials Protection and Performance (AMPP)
Web: ampp.org
Previously NACE and SSPC. Standard Recommended Practices, reports, conference proceedings, Materials Performance Journal, Corrosion Journal, training and certification courses in corrosion

Transportation Research Board (TRB)
Web: www.trb.org
Annual Conference, National Cooperative Highway Research Program conducts corrosion and concrete repair related research, NCHRP and Strategic Highway Research Program (SHRP) reports available online through website. Transportation Research Record annual publication of selected conference papers

RILEM
Web: www.rilem.org
Publishes reports and monthly Materials and Structures – Matériaux et Constructions Journal

WTA e.V.
Web: www.wta.de

Wissenschaftlich-Technische Arbeitsgemeinschaft für Bauwerkserhaltung und Denkmalpflege, or International Association for Science and Technology of Building Maintenance and Monuments Preservation. German language conferences, seminars, expert discussions, proceedings, books and recommendations. WTA publishes the International Journal for Technology and Applications in Building Maintenance and Monument Preservation (which is part of the membership package).

Glossary and index

These definitions are not full, accurate scientific or dictionary definitions and may be incomplete if used outside the context of the subject of corrosion of steel in concrete.

acid   A solution that (among other things) attacks steel and other metals and reacts with alkalis, forming a neutral product (a salt) and water. Acids are also used to dissolve concrete for chemical analysis.

alkali   A solution that (among other things) protects steel and other metals from corrosion and reacts with acids, forming a salt and water.

anode

1. The site of corrosion in an aqueous corrosion cell (a combination of anodes and cathodes).
2. An external component introduced into a cathodic protection system to be the site of the oxidation (q.v.) reaction and prevent corrosion of the metal object to be protected.
3. The positive pole of a simple electrical cell (battery).

binding (of chlorides *see also* chloroaluminates) 3, 24, **29–30**, 59

bond strength
coating 162, 182, 192
membranes 129
reinforcement 124–5, 202–3
repairs **116–17, 121–2,** 124, 133, **138,** 232
sprayed concrete **138, 168–70**

carbonation The process by which carbon dioxide (CO_2) in the atmosphere reacts with water in concrete pores to form carbonic acid and then reacts with the alkalis (q.v.) in the pores, neutralizing them. This can then lead to the corrosion of the reinforcing steel.
carbonation 2, 8, 11, **17–21,** 23, 26, 29, 30, 35, 36, 42, 49, 50, 51, 54, **55–7,** 59, 75–6, 116, 122, **125–7,** 135, 138, 186, **204–7,** 215, **219–24,** 225, 231, **237–40, 243–4,** 251, **256–7,** 271–2, 275–6

cathode
1. The site of a charge balancing reaction in a corrosion cell.
2. The protected metal structure in a cathodic protection system.
3. The negative pole of a simple electrical cell (battery).

cathodic protection A process of protecting a metal object or structure from corrosion by the installation of a sacrificial anode (q.v.) or impressed current system that makes the protected object a cathode (q.v.) and thus resistant to corrosion.
criteria **146, 182–8,** 192–4, 227

cathodic protection anode A cathodic protection anode for steel in concrete can be a conductive paint or other conductive material that will adhere to concrete, or a metal mesh or other conductive material that can be embedded in a concrete overlay on the surface or in chases or holes cut in the concrete of the structure to be protected. Anodes may be impressed current or galvanic.
cathodic protection
anode (*see* anode)

cement (paste) Portland cement is a mixture of alumina, silica, lime, iron oxide and magnesia ground to a fine powder, burned in a kiln and ground again. Cement paste is the binding agent for mortar and (Portland cement) concrete after hydration.

chloride The negative ion (q.v.) in salt (sodium chloride), found in sea salt, deicing salt and calcium chloride admixture for concrete. Chloride ions promote corrosion of steel in concrete but are not used up by the process so they can concentrate and accelerate corrosion.

chloroaluminates Chemical compounds formed in concrete when chlorides combine with the tricalcium aluminates (C_3A) in the hardened cement paste. These chlorides are no longer available to cause corrosion. Sulphate resisting cements have a low C_3A content and are more prone to chloride-induced corrosion than normal Portland cement based concretes.
chloroaluminates (*see also* binding) 22, 30, **58–9**

coating (for concrete) 7–8, **125–8, 220–5,** 227–8, **230–1,** 232–3, 257, 270–2, 276

concrete Ordinary Portland cement concrete is a mixture of cement (q.v.), fine and coarse aggregates and water. The water reacts with the cement to bind the aggregates together.

pH   A measure of acidity and alkalinity based on the fact that the concentration of hydrogen ions [H~} (acidity) times hydroxyl ions [0H1 (alkalinity) is 10W' moles/i in aqueous solutions:

$$[H^+] \times [OH] = 1 \times 10^{-14}$$

$$pH = - \log[H^+]$$

$$pH + pOH = 14$$

i.e. a strong acid has pH = 1 (or less), a strong alkali has pH = 14 (or more), a neutral solution has pH = 7. Concrete has a pH of 12 to 13. Steel corrodes at pH 10 to 11.

pore (water)   Concrete contains microscopic pores. These contain alkaline oxides and hydroxides of sodium, potassium and calcium. Water will move in and out of the concrete saturating, part filling and drying out the pores according to the external environments. The alkaline pore water sustains the passive layer if not attacked by carbonation or chlorides (q.v.).

post-tensioning   is a form of prestressing concrete by casting ducts and/or anchors into the concrete. After the concrete has hardened and developed sufficient strength, cables in the ducts or attached to the anchors are tensioned and secured to the anchors. Ducts may be filled with cement grout (bonded) or with protective grease (unbonded). The grouting process is vital to the corrosion protection of the post-tensioning cables.

prestressing   The process of applying compressive stress to concrete using steel rods, wires or bars to make stronger elements than by conventional reinforcement. There are two forms of prestressing; pretensioned and post-tensioned. Pretensioned structures and elements are made by applying a load to the steel, casting the concrete around it and then, when it has hardened and developed sufficient strength, releasing the load on the steel which is then taken up by the concrete.

reduction   Chemically this is the reverse of oxidation (q.v.). The incorporation of electrons into a nonmetal oxidizing agent when a metal is oxidized. When oxygen (O_2) oxidizes iron (Fe) to Fe^{2+} it receives the electrons that the iron gives up and is itself reduced:

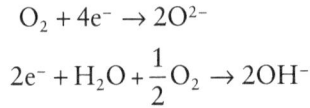

$$O_2 + 4e^- \rightarrow 2O^{2-}$$

$$2e^- + H_2O + \frac{1}{2}O_2 \rightarrow 2OH^-$$

are reduction reactions.

reference electrode   Usually, a pure metal in a solution of (fixed) concentration. The half reaction of the metal ions dissolving and reprecipitating creates a steady potential when linked to another half cell. Two half cells make an electrochemical cell that can be a model for corrosion or be used to generate electricity (a single cell, often called a battery). Reference half cells are connected to reinforcing steel to measure 'corrosion potentials' that show the corrosion condition of the steel in the concrete.

reinforced concrete   Concrete containing a network of reinforcing steel bars to make a composite material that is strong in tension as well as in compression. Smaller volumes of material can therefore be used to make beams, bridge spans, etc. compared with unreinforced concrete, brick or masonry.

rust   The corrosion product of iron and steel in normal atmospheric conditions. Chemically it is hydrated ferric oxide $Fe_2O_3 \cdot H_2O$. It has a volume several times that of the iron that was consumed to produce it.

sacrificial anode cathodic protection (*see* galvanic cathodic protection)

SHRP   The Strategic Highway Research Program. A $150 million research effort that spent about $10 million on corrosion of reinforced concrete bridges suffering from chloride-induced corrosion. SHRP produced about 40 reports covering assessment, repair and rehabilitation methodology. All reports can be downloaded from www.trb.org

steel   An alloy of iron with up to 1.7% carbon to enhance its physical properties.

For Product Safety Concerns and Information please contact our EU
representative GPSR@taylorandfrancis.com
Taylor & Francis Verlag GmbH, Kaufingerstraße 24, 80331 München, Germany